Routledge Introductions to Development

Series Editors:
John Bale and David Drakakis-Smith

Population and Development in the Third World

T0173935

In the same series

John Cole
Development and Underdevelopment
A profile of the Third World

David Drakakis-Smith
The Third World City

Avijit Gupta
Ecology and Development in the Third World

John Lea
Tourism and Development in the Third World

John Soussan
Primary Resources and Energy in the Third World

Chris Dixon
Rural Development in the Third World

Alan Gilbert
Latin America

Janet Henshall Momsen
Women and Development in the Third World

David Drakakis-Smith
Pacific Asia

Rajesh Chandra
Industrialization and Development in the Third World

Tony Binns
Tropical Africa

Jennifer A. Elliott
An Introduction to Sustainable Development

Allan Findlay
The Arab World

Mike Parnwell
Population Movements and the Third World

George Cho
Trade, Aid and Global Interdependence

David Simon
Transport and Development in the Third World

Allan Findlay and Anne Findlay

Population and Development in the Third World

Routledge
Taylor & Francis Group

LONDON AND NEW YORK

To our parents

First published in 1987 by
Methuen & Co. Ltd

Reprinted 1991, 1994, 1995, 2000
by Routledge
2 Park Square, Milton Park, Abingdon, Oxon OX14 4RN
711 Third Avenue, New York NY 10017

Routledge is an imprint of the Taylor & Francis Group

© 1987 Allan M. Findlay and Anne M. Findlay

Typeset by Hope Services (Abingdon) Ltd.

British Library Cataloguing in Publication Data
Findlay, Allan M.
Population and development in the Third World.—(Routledge introductions
to development)
1. Developing countries—Population
2. Developing countries—Economic conditions
I. Title   II. Findlay, Anne M.
304.6'2   HB884

Library of Congress Cataloguing in Publication Data
Findlay, Allan M.
Population and development in the Third World.
(Routledge introductions to development)
Includes index.
1. Developing countries—Population. 2. Developing countries—Population policy.
I. Findlay, A. M. II. Title. III. Series.
HB884.F56 1987   304.6'09172'4.   86–31094

ISBN: 978-0-415-06584-9

Contents

Acknowledgements

The authors would like to thank the following for kind permission to reproduce material in this book: Butterworth Scientific for figure D.1, J. Clarke for material used in chapter 3, J. Findlay for plates 1.1 and 1.2, C. Hughes for plate 5.1, and Reidel Publications for figures E.1 and E.2. We owe a special debt of gratitude to our colleague John Jowett who not only contributed figure 4.1 and plates 6.1, 6.2 and 6.3 but also generously agreed to write two case studies on India and China to illustrate chapters 4 and 6 respectively. We are also most grateful to M. Shand, Y. Wilson and L. Hill who prepared the illustrations for this volume and to B. Bolt who typed the final manuscript.

Preface

The nature of the relationship between population and development has long been disputed. The early 1970s saw the publication of several books such as the *Population Bomb*, written by the American biologist Paul Ehrlich, which warned of the danger that the human race would breed itself into a catastrophic crisis situation in which the finite physical resources of the earth would no longer be able to support the world's ever growing human population. It was in an atmosphere of extreme concern about the possible detrimental effects on development of rapid population growth that a World Population Conference was organized in Bucharest in 1974. No less than 136 countries agreed at this conference to a World Population Plan of Action which called for increased efforts, particularly in the less developed countries, to introduce family planning programmes and to reduce rates of population growth in order to conserve resources and improve standards of living. In 1976 the United Nations International Fund for Agricultural Development was established in response to claims that world population growth had reduced world food stocks to their lowest levels since the Second World War. At about the same time reports to the experts of the self-styled Club of Rome announced that mass starvation could be avoided only through radical changes in population growth rates and in methods of resource utilization.

By contrast with these pessimistic views of population and development trends the 1980s saw the production of foodstuffs in the more developed economies far outstripping demand. In 1984 the United Nations World Food Council suggested that many of the predictions made in the early

1970s were wrong and that for the foreseeable future the world was well able to produce all the food it would need for its growing population. In 1985 the United Nations Food and Agricultural Organization reported on the remarkable changes which had occurred in cereal production as a result of the introduction of new high-yielding strains. These have transformed world cereal production to achieve levels inconceivable a decade earlier and have created problems of surplus storage rather than of deficits. Despite these advances, the greater optimism of the 1980s has been balanced by continued reports of drought, famine, deforestation and soil erosion in many of the poorer countries of the world. Although some sources continue to point to rapid population growth as the main culprit for undernourishment and hunger, most authorities suggest that inadequate development efforts and a lack of resource management are more likely sources of the miseries faced by the populations of many Third World countries.

It was therefore in a quite different atmosphere and with a different outlook that the second World Population Conference was held in Mexico in 1984, precisely a decade after the Bucharest meeting. The proposals arising from the conference reflected an increasing awareness that the success of population policies was dependent on their being set within an appropriate framework for social and economic development. Development policies, plans and programmes should reflect the inextricable links between population, physical resources, the environment and the standard of living.

The keynotes of the so-called *Declaration of Mexico*, signed as a result of the second World Population Conference, provide an interesting summary of the officially accepted views of the relationship between population and development. The Declaration emphasized that, while family-planning programmes should be sustained, priority should be given to integrating population and development factors through schemes which take fully into account the need for a more rational utilization of natural resources and the need to protect the physical environment. Population programmes were to reach beyond the goal of fertility reduction by seeking to improve the status of women and enhance their roles in the family, the community and society. To achieve this, institutional, economic and cultural barriers to fuller female participation in the economy and in society would need to be removed.

The *Declaration of Mexico* also pointed to the fact that an increasing proportion of the world's population live in very large cities, and that by the end of the century an estimated 48 per cent would be urban residents. Consequently population policies should pay greater attention to the different needs of rural and urban areas and be closely integrated with rural and urban development strategies. Population policies should be based on a fuller evaluation of the costs and benefits to the individuals, groups and regions involved and should respect basic human rights by seeking to use incentives rather than restrictive measures.

The principal themes

This book seeks to examine some of these issues in greater detail. It commences with an investigation of the nature of population growth in the less developed countries and explores how this influences prospects for economic development. The term 'population' is interpreted in the broadest sense to include not only demographic processes but also the contribution of population to the labour market and the role of people as consumers of natural resources in influencing development. The relationship between population and development is therefore highly complex and involves two-way interactions, with populations both affecting and being affected by their economic, social and political environments. The first part of this book investigates the influence of development on population while the remainder places greater emphasis on the role that population factors play in the development process.

Figure 1, 'Population needs and population change', represents some of the interactions studied in this book. Chapter 1 examines the characteristics of populations in the less developed world and asks why they have grown so extremely rapidly during the twentieth century. Chapter 2 investigates the economic, social and political factors which affect natural population change with particular reference to the ways in which economic development influences family size and fertility levels. Chapter 3 attempts to offer a time perspective on how trends in economic development appear to have been associated with changes in demographic processes such as birth rates and death rates, while chapter 4 examines the inverse relationship between population growth and the effects of this growth on the economy of developing countries through the food needs of the population. Since populations have many non-food needs such as shelter and clothing, as well as a desire for many manufactured products, population growth also influences the non-food sector of developing economies (chapter 5), with people acting both as producers and consumers of industrial goods and services. Both chapters 4 and 5 emphasize the mobility of populations which results from regional inequalities in resource availability and in people's perception of the availability of opportunities to meet their food and non-food needs. Population therefore affects economic development, not only through natural demographic change but also through the impact of migration in redistributing people between regions and indeed between nations. The final chapter examines population policies, through which governments attempt to change the size and composition of their populations. It is, of course, true that all forms of government intervention in an economy affect patterns of natural increase and patterns of migration, and one of the greatest difficulties in evaluating planned population policies is to identify whether demographic changes occur as a result of these policies or

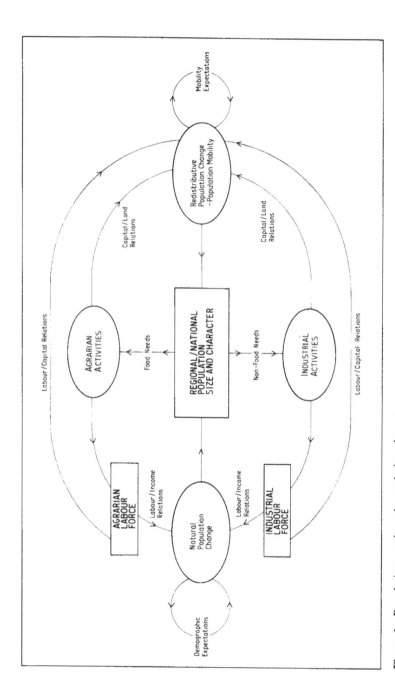

Figure 1 Population needs and population change

because of more general economic and social forces. Sadly some governments continue to publicize manifestos which suggest that population growth and economic growth are irreconcilable alternatives, whereas there is growing evidence to show that the development of population resources is a critical step towards achieving sustained economic development.

1
Population growth

Population growth and standards of living

For the greater part of human history, there were only a few million people inhabiting the earth. This is not surprising in view of the limited ability of hunting and gathering economies to meet their food needs. The introduction of methods of settled cultivation increased the capacity of the world to support mankind, and the earth's total population is believed to have grown to around 300 million by the time of Christ's birth. The really dramatic changes in the size of the world's population began to occur in the eighteenth and nineteenth centuries. Although different views have been propounded as to the precise causes of these changes, there can be little doubt that the arrival of industrial production in a small number of countries had profound implications for the ways their societies and economies were organized and would develop. Rapid population growth occurred first, therefore, in places associated with early industrial developments. By 1850 the human population of the earth had risen to about 1000 million persons, and within a century the figure was to increase to 2500 million. Not only were these population increases happening at an ever growing pace, but by the twentieth century it became clear that there was no longer a direct geographical association between places of industrial development and places of population growth, since it was increasingly the less developed countries of the world which were experiencing the largest increases in population. Between 1950 and 1980 the rate of population growth was most rapid in Latin America and Africa whose populations more than doubled during this short period, but

Table 1.1   Populations of the less developed world (millions)

	1950	1980	Population increase 1950–80
Latin America	164	364	200
Africa	220	470	250
East Asia	673	1,175	502
South Asia	716	1,404	688

population increases were even larger in absolute terms in South and East Asia (table 1.1). By 1980 the world had 4800 million inhabitants and there was the prospect that a further 2000 million would be added to the total by the end of the century. Compared with the long course of human history, the nineteenth and twentieth centuries have therefore seen an astounding population explosion. This has both influenced and been influenced by patterns of economic and social development.

At the level of individual countries population growth occurs because of either a net surplus of births over deaths (termed natural increase) or from net gains by migration. Historically, natural population change has been the major force accounting for national increases and decreases in population numbers. The very rapid population growth evident in table 1.1 has occurred mainly because there have been substantial reductions in the crude death rates (number of deaths per 1000 persons) of the less developed countries in recent decades, a phenomenon that has not been accompanied by comparable reductions in crude birth rates (number of births per 1000 persons). The trend has been common to most African, Asian and Latin American countries, but in a small number of cases migration has proved a more important force in influencing the total population size. For example between 1973 and 1982 the two countries in the world with the highest population growth rates were the United Arab Emirates and Kuwait. Both experienced population growth predominantly as a result of labour immigration from other countries and had more immigrants than native-born in the male population.

Population growth rates are measured by calculating the average annual percentage change in a country's population. The most striking feature of the world pattern of recent national population growth is the geographical concentration of very high growth rates in Africa, Latin America and South and East Asia (figure 1.1). By comparison the more industrialized economies of the northern hemisphere, as well as of Australasia, had much lower rates of population growth. Between 1973 and 1982 thirty countries, all of them in the less developed world, sustained annual population growth

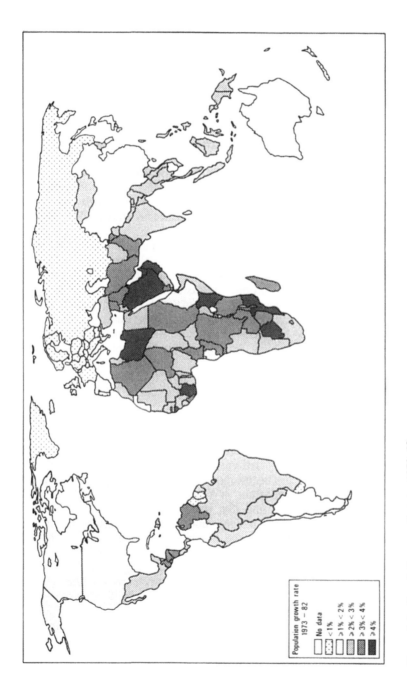

Figure 1.1 Population growth rates 1973–82
Data source: The World Bank Atlas, 1985

rates of over 3 per cent per annum. By contrast the more developed countries often had population growth rates of less than 1 per cent per annum.

Since this book is particularly concerned with the relationships between population and development it is important to define what is meant by 'development' before proceeding further. No single measure can really hope to define completely what is meant by the level of 'development' in a particular country. Terms such as 'primitive', 'backward' and 'underdeveloped' have tended to give way to descriptions such as 'less developed' and 'Third World'. This change has occurred in response to the realization that development is a complex political, social and economic phenomenon, and that many of the early studies of development were strongly biased by European and American cultural values and incorrectly equated development with a country becoming more 'western'.

Development has most frequently been assessed in economic terms using an indicator called the Gross National Product (GNP). This is defined as the value of total output of goods and services accruing to the inhabitants of a country. The gross national product of a country ignores many important social aspects of development such as nutrition levels, education, health care and life expectancy, all of which contribute to the quality of life. It also fails to represent significant economic attributes such as income distribution. Nevertheless GNP provides an interesting yardstick of the volume of economic activity occurring within a country. In terms of figure 1.2 on page 5, such a measure is rather useful in providing a broad indication of the ability of different countries to provide a living for their populations from their agricultural, industrial and service economies. This ability will, of course, change through time as a result of access to higher levels of technology and will vary regionally depending upon the way that people and resources are organized within specific political and social contexts. This organizational element is in itself very important in influencing population change. For example, it has been suggested that a society which encourages women to become an important part in the 'modern' argicultural and industrial labour force will find that many women accept this role and choose to work rather than having children, thus delaying the age of marriage and reducing the average family size. There are a number of reasons why this relationship is hard to test in practice as will be shown in chapter 2. This is only one of many examples of the ways in which the social organization of economic activities and population change are interrelated, but it demonstrates the *indirect* and *complex* nature of the links between economic and demographic processes.

The most commonly studied measure of population and development is the ratio of a nation's gross national product to its population (GNP per capita). The pattern of GNP per capita in 1982 is mapped in figure 1.2. It can

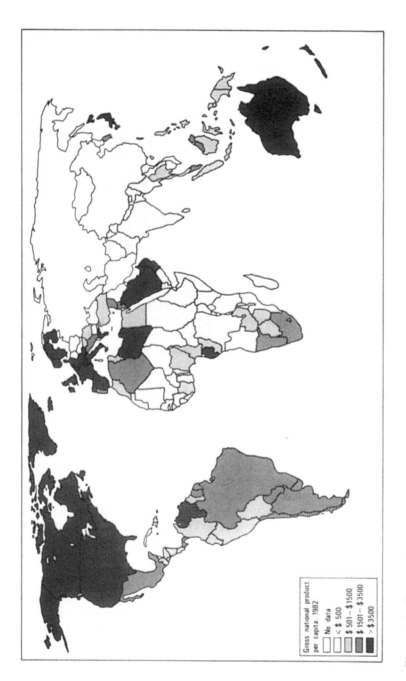

Figure 1.2 Gross national product per capita 1982
Data source: The World Bank Atlas. 1985

Gross national product
per capita. 1982

No data

< $ 500

$ 501– $1500

$ 1501– $3500

> $3500

be seen that the wealthiest populations in the world are geographically concentrated on either side of the North Atlantic as well as including Australia, New Zealand and Japan. Evidence that wealth and development are not entirely synonymous is evident from the presence of the leading oil exporters of the Arab world amongst the highest-income countries, yet on nearly every other indicator these countries would not appear to be 'developed'.

The immense disparity in the distribution of wealth relative to population is emphasized by statistics produced by the United Nations which indicate that in 1982 nearly half the world's population lived in the least developed countries with an average GNP per capita of only $270, while at the other extreme a sixth of the world's population lived in countries with an average GNP per capita more than forty times as great (figure 1.2). These statistical indicators, crude as they are, reinforce the need to investigate the relationships which exist between population and development, in order to discover whether rapid population growth is inherently associated with low levels of economic development.

Why do Third World countries have much higher population growth rates than the more developed countries? This profound question can be answered at many different levels. To state the direct demographic answer, that high rates occur in the Third World because the number of births far exceed the number of deaths while in the more developed countries the two processes are almost evenly matched, leads only to the question 'Why do demographic differences occur in this way?' While this question is tackled in more detail in the next chapter, it is useful to note at this point that high population growth rates are themselves a cause of further rapid population growth. Why should this be the case? Rapid population growth can be self-perpetuating for the simple reason that the number of births in a population is itself a function of how many young women there are of child-bearing age. The high population growth rates of the less developed countries have meant that these countries also have a very high proportion of young women in their populations and therefore a very great potential for future population growth. The 'momentum' created by high fertility and declining mortality means that the number of women entering the child-bearing ages in many countries represents an ever larger proportion of the total population. Even if the number of births per woman declines the birth rate (number of births per thousand persons) will remain high and the absolute number of births will be greater than previously.

The peak rate of population growth in the less developed countries was in fact reached in the mid-1960s and since then the rate of growth has fallen slightly. This has been in part due to the reduction of the number of children born to each woman in China, the world's largest population unit, and it should be noted that population growth rates remain extremely high in most

Table 1.2   Rates of population growth and national wealth

Population growth rate, 1973–82	Number of countries	Total population, 1982 (millions)	Mean GNP per capita (US $ 1982)
Less than 1%	33	346.1	9,310
1% to less than 2%	40	1,618.9	3,400
2% to less than 3%	49	1,704.2	760
3% and more	37	324.1	1,590
No data	30	574.7	n.a.

Source: The World Bank Atlas (1985).

other Asian and African countries. As a result of the demographic 'momentum' described above, the number of people being added to the populations of the less developed countries by the processes of natural increase is therefore higher in absolute terms than ever before, despite a slight slackening in rates of growth. For example, between 1965 and 1980 Brazil's total fertility rate (see chapter 2 for a definition) fell from 5.8 to 4.0, a decline of almost 30 per cent, but the total number of births increased from about 2.9 million per year in the late 1950s to about 3.7 million per year in the early 1980s. Although Brazil's total fertility rate is expected to continue to fall to the point where, by 2025, the population is only just reproducing itself, it is estimated that by then the number of births per year will be 3.9 million. Not only are the populations of many countries like Brazil larger than ever before, but, as a comparison of figures 1.1 and 1.2 shows, the largest absolute gains have been made in the poorest countries.

The inverse relationship between standard of living (as measured by GNP per capita) and population growth is not, however, a very precise one (table 1.2). This is the case partly because of the inadequacy of GNP per capita as an indicator of standard of living and partly because the relationships between economic and demographic growth processes are indirect rather than direct. For example, the measurement of GNP per capita assumes that national economic wealth is evenly distributed between all members of a population, a situation which does not occur in every country and one from which many countries deviate to a considerable extent as a result of their political and social structures. A more important reason for the imperfect association between GNP per capita and rates of population growth is that economic factors are only one of many forces influencing demographic trends. This is particularly true with regard to trends in family size and crude birth rates. For example, Bolivia, Thailand and Zimbabwe all had very similar levels of GNP per capita in 1978, yet in Thailand the crude birth rate was only 32 per thousand, while in Bolivia it was 44 per thousand and in

Zimbabwe 49 per thousand. This diversity in fertility behaviour becomes even more remarkable when countries within the same culture region are compared.

National population structures

Although a nation's income level does not have a direct influence on its population structure, demographic forces do have very clear economic and political implications. For example, the high population growth rates of less developed countries have increased the need for investment in educational facilities as well as requiring an expansion of job opportunities for the ever increasing number of young persons. Figure 1.3 illustrates the age structures of several developing countries and contrasts these with a similar diagram for the United Kingdom. These age–sex diagrams are often referred to as 'population pyramids'. The vertical axis of each pyramid is divided into five-year age bands, while the horizontal axis shows the percentage of males and females in the total population belonging to each 'cohort'. Age–sex pyramids are a useful tool for examining a nation's population characteristics, since the structure reflects the long-term trends of the country's fertility and mortality patterns, as well as indicating the shorter-term impacts of wars, epidemics and migrations.

As can be seen from figure 1.3 the term 'pyramid' is only really applicable to the less developed countries, and the age–sex structures of countries like Britain show a much more stable population situation with moderately little change in the proportion of the population belonging to different age cohorts under the age of 60. By comparison, the developing countries have a tapering age structure that merits the use of the term pyramid. This shape occurs on the one hand because of the high crude birth rates and consequently the large number of children present, and on the other hand because of the higher death rates which result in each cohort becoming progressively smaller as it ages. The broad base of the population pyramids of Third World countries is seen at its most extreme in the case of the Kenyan pyramid, which shows that no less than 20.9 per cent of its population was under five years of age in 1983. In Sudan a figure of 18.5 per cent was recorded. Not all developing countries have as broad a base as these African states, but it would be fair to say that the majority have extremely youthful populations. This position is being maintained, despite the influence of improved hygiene and better medical facilities in reducing death rates, because of sustained high birth rates.

Figure 1.3 shows that in most countries there is not a perfect correspondence between the number of men and women. Although this asymmetry is not very marked, in nearly all countries there is a slight excess of males over females at birth. For example, in the youngest Indian cohort (0–5 years) in

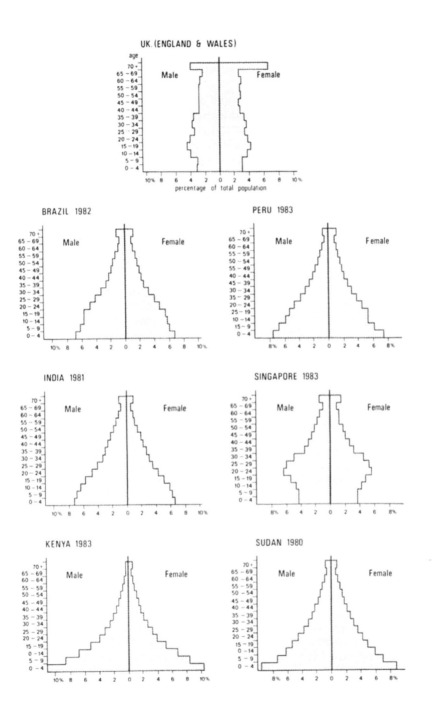

Figure 1.3   Comparative national population structures

1981 there were 47,791,000 males and only 45,364,000 females (or 7.1 per cent and 6.7 per cent of the population respectively). This male preponderance at birth apparently reflects a natural adaptation to higher pre-natal and infant mortality rates amongst males. Although sex ratios at birth are influenced by many other factors, including the desire for male children and the admission of female infanticide in some Asian countries, there can be little doubt that the degree of pre-natal hygiene and care has an important influence on sex ratios at birth. The pyramids of figure 1.3 show that with increasing age the differential between the numbers of males and females generally decreases, although in some countries, such as India where the average age of marriage remains extremely low, high levels of maternal mortality are recorded and account in part for the substantial excess of men in the 15–19 age cohort.

In the past the age structures of many Third World countries resembled those of India and Peru (figure 1.3). Today there is evidence of increasing divergence from the characteristics of these tapered pyramids. Consider once again the case of Brazil. In 1950 Brazil had a very broad base just like that of Kenya today, but by 1982 as a result of falling birth rates the percentage of the population under five years of age had been reduced to only 13.8 per cent. This nevertheless remains the country's largest single age cohort. The slight reduction in fertility rates evident in the marginally more developed countries of the Third World, such as Brazil, gives their age–sex structures a 'beehive' rather than a 'pyramid' shape. In Singapore the effect of falling crude birth rates, which have been halved since the 1950s, is even more marked, resulting in an increase in the average age of the population and a marked pinching of the base of the population pyramid as a result of an absolute reduction in the number of births. It should be noted that this structural change has occurred because of modifications in fertility behaviour rather than because of mortality factors.

Although the effect of migration is not very evident in any of the national population pyramids of figure 1.3, this process can also operate in an age- and sex-selective fashion. Historically, it has been men rather than women who have been the dominant international migrants, and this can have a significant effect in reducing the number of men in the most active cohorts of countries of emigration and inversely in increasing the male population of countries of immigration. In Singapore the cohorts of males between 20 and 40 years of age are swollen by the presence of some 100,000 immigrant workers from other countries who account for no less than 12 per cent of the total workforce. Migration as a sex-selective process has a more marked effect on population structures at the scale of regions and cities. At these scales migration is not necessarily biased towards men. For example, in Latin America and parts of Eastern Asia it is women who predominate in internal labour migration from rural to urban areas.

Spatial dimensions of population growth

Demographic statistics are most readily available at a national scale and it is probably for this reason that population growth is usually analysed on this basis. It is pertinent to ask whether there are any other reasons for accepting the state as an appropriate scale at which to examine patterns of population growth and development. A case can be made for study of population change at this scale, since it is at the state level that political bodies and institutional forces are often organized, and it is at this scale that economic forces often impinge on demographic processes. For example, where two neighbouring states have followed very different policies on family planning and birth control, as in the cases of Tunisia and Algeria, the result can be that international frontiers become demographic divides.

In many other instances, however, state policies are much less important in influencing patterns of population growth than many other forces operating at regional or perhaps international levels. Many population groups with distinct cultural or ethnic bonds find themselves geographically divided by international frontiers. For example, the Kurds have no identity with a unique political state but find their community divided by the borders of Iran, Iraq and Turkey. The political map of Africa results in many similar examples of so-called 'national' boundaries which were drawn up by former colonial rulers and which in fact cut across traditional tribal areas. One finds not only the artificial division of homogeneous population units, but also the presence of states lacking any genuine 'national' identity, and consisting instead of many minority populations, each with different demographic and cultural traditions. Plural societies (societies with many different population groups holding separate identities) are found in many Third World states, yet information about the demographic characteristics of minority populations is scanty.

Detailed spatial analysis of population processes is important, not only because birth and death rates may vary between different ethnic groups living within a state, but also because population migration between regions can lead to very significant spatial variations in population growth within the state. In Third World countries the regions with the highest population growth rates have usually been those containing the largest urban centres. By contrast rural areas with low standards of living and poor prospects for economic improvement often experience minimal population growth since out-migration reduces the effects of additions to the population due to natural increase.

Migration to cities was responsible for very high urban growth rates in the 1950s in Latin America and East Asia, but in Africa and South Asia it was only in the 1970s and early 1980s that migration had its strongest impact on urban areas. Table 1.3 shows that, as a consequence of population

Table 1.3 Total population, percentage urban and rural and urban growth rates of selected countries

	1980 population Number (millions)	% urban	Average annual growth rates (%), 1975–1980 Total	Urban	Rural
Bangladesh	88.2	11.2	2.8	6.7	2.4
Brazil	122.3	67.0	2.4	4.0	−0.5
Colombia	25.8	70.2	2.1	3.5	−0.8
Cuba	9.7	65.4	0.8	1.7	−0.6
Egypt	42.0	45.4	2.6	3.4	1.9
Ethiopia	31.5	14.5	1.8	6.1	1.2
Guatemala	7.3	38.9	3.0	4.0	2.4
Indonesia	148.0	20.2	1.7	3.6	1.3
Jordan	3.2	56.3	3.7	4.9	2.2
Kenya	16.5	14.2	4.0	7.3	3.5
Mexico	69.8	66.7	3.0	4.1	0.9
Philippines	49.2	36.2	2.7	3.8	2.1
Sudan	18.4	24.8	2.7	6.7	1.6
Venezuela	15.6	83.3	3.5	4.3	0.1
Zaire	28.3	39.5	2.8	5.3	1.3

Source: adapted from *Population Reports* (1983) p. 248.

redistribution, average population growth rates in urban areas far exceeded rural population growth. Indeed out-migration from rural areas actually resulted in net rural population decline in some countries in Latin America.

One of the curious paradoxes of the so-called 'population explosion' of Third World cities is that on the one hand it is a consequence of the substantial spatial inequities in welfare which exist within countries between urban and rural areas, while on the other hand these very same cities offer minimal opportunity for betterment to many of the in-migrants who come to them in search of employment and services. The complicated relationships between migration, urban growth and economic development are taken up in greater detail in a later chapter (and also in the books by Hugo and Drakakis-Smith in this series), but it is worth noting at this stage that most research studies show that migration to the cities of Third World countries accentuates existing problems, adding to urban unemployment and under-employment, increasing pressure on inadequate housing resources and augmenting societal and psychological stresses amongst urban populations. Only if these serious problems can be overcome will it be possible to interpret the trend towards population concentration in Third World cities as a beneficial consequence of economic and social development.

This chapter has examined, at a number of scales, several different dimensions of population growth and development. In the chapters that follow the components of population growth are examined in more detail, prior to examining the impact of demographic change on economic development.

Key ideas

1 The highest rates of population growth are concentrated geographically in the Third World.
2 The relationship between population and development is two-way with population characteristics influencing economic development as well as development affecting demographic trends. The nature of the relationship is highly complex with most influences being indirect.
3 Gross national product per capita is a common indicator of economic development, but it is only of limited value in development studies.
4 Analysis of national population structures shows that, even when population growth rates slow down, substantial absolute population gains may continue to occur.

2
Mortality and fertility levels in the Third World

Measuring mortality and fertility

After briefly exploring in the previous chapter the complex relationships which exist between rates of population growth and economic development, it is now appropriate to examine in more detail the factors which result in natural population increase: namely mortality and fertility.

Before any explanation of mortality and fertility patterns in less developed countries can be attempted, it is necessary to consider carefully how trends in deaths and births are measured. The most rudimentary measures are termed 'crude death rates' and 'crude birth rates' and measure the ratio of deaths and births in a year to the total population. These indices, while widely used and easily calculated, are of limited value since they ignore the composition of a population, and, as has been shown in the previous chapter, the age and sex characteristics of a population strongly influence demographic processes. For example, in 1978 the crude death rate in less developed countries with a high proportion of young people such as Colombia or Paraguay were of the order of 8 per thousand, while countries with more elderly populations such as the UK and West Germany had rates of 12 per thousand. Consequently, in making international comparisons of birth rates and death rates, more refined demographic measures are normally used in order to in some way take account of population composition.

In studying mortality trends two measures which achieve this are infant mortality rates and life expectancy. Mortality rates calculated for specific population groups by age cohort or sex overcome most of the problems of aggregate crude mortality rates. The infant mortality rate is a particularly informative indicator since many deaths are concentrated in the first year of life. This measure is therefore a useful yardstick for comparing health conditions in developing countries. The infant mortality rate is conventionally defined as the number of deaths in a particular year of children under one year of age per thousand live births. Life expectancy, paradoxical as it may seem, is a most useful measure of mortality since it estimates the average number of years that a cohort of people born in a particular year are likely to live, given the age-specific death rates prevailing at the time. Similarly, with fertility rates, variations in the crude birth rate reflect to some extent spatial variations in population structures and include the large mass of men, children and elderly people not involved in child bearing. Consequently measures such as the total fertility rate are more valuable indicators of fertility behaviour. The total fertility rate (given certain assumptions) indicates the number of children that would be born to a thousand women passing through the child-bearing age cohorts.

Mortality patterns and life expectancy

There is a broad geographical association in the less developed countries between high age-specific mortality rates, high infant mortality rates and low life expectancy. It would however be erroneous to assume that mortality levels or life expectancy are a direct function of standards of living. This is easily shown in the case of many less developed countries which have experienced over the last thirty years very slow, and in some cases negative, rates of economic growth, yet have recorded substantial declines in mortality rates and gains in the average life expectancy of the population at birth.

The most striking feature to emerge from a detailed investigation of patterns of mortality in the less developed countries is the geographical diversity rather than the uniformity of mortality rates. Some demographers suggest that, because certain infectious diseases are either endemic or more virulent in tropical climates, mortality rates have always been higher in equatorial countries than in other parts of the world. While the causes of the decline in death rates and the rise in life expectancy in the industrialized countries has been much debated (being attributed by some to improved nutrition with rising living standards, by others to medical advances and developments in public health provision and by yet others to the changing internal character of infectious diseases) the increase in life expectancies in the less developed countries has often been interpreted as being dominantly a result of their level of contact with western nations.

Until this century the extremely low life expectancies in most less developed countries were sustained by periodic catastrophic epidemics and famines. For example, the French demographer Lardinois has shown how the combined effects of famine and a cholera epidemic in southern India in the years 1876–78 not only increased the number of deaths to ten times the normal level but also reduced the number of conceptions by about half. The activities of the World Health Organisation and other international agencies did much to help eradicate and reduce the effects of diseases like smallpox and malaria which once accounted for the early deaths of millions of people. Particularly effective were the malaria-eradication programmes launched in the 1950s and the international programme to reduce the effects of smallpox in the late 1960s.

The rapidly rising life expectancies of the populations of many less developed countries in the 1950s and 1960s have been dominantly attributed to the introduction from the west of medical and health programmes. Life expectancies rose earlier in Latin America than in India. In both cases there was a surge in life expectancies in the post-war period. In India, for example, life expectancy at birth was still only 25 years until the 1920s and had only just reached 30 by 1940. By 1960 it was over 40 and by 1982 it had risen to 55 years. While immunization and vector control were undoubtedly important in fighting infectious diseases such as malaria, tuberculosis and smallpox, significant progress in combatting disease and ill health in the less developed countries has also been partly the result of economic development factors such as improvements in infrastructure (better water supplies and improved sewage systems). Indeed international health organizations had predicted that life expectancies in less developed countries would rise much more rapidly in the 1970s than in fact occurred.

The main reasons for the shortfall seem to be the persistence of mass poverty in the Third World, and with it the continuation of many diseases linked to poor nutrition, poor sewerage and inadequate and polluted water supplies. At a national level considerable differences in life expectancies still exist betwen the countries of the Third World, with the average for tropical South America being 62 years and temperate South America achieving 68 years, while in South-east Asia life expectancy remains at 53 years and in equatorial Africa it is only 45 years.

Spatial variations in life expectancies are also evident at a more detailed scale of analysis, for example, between different regions of a country or even between different districts of a city. Recent research has increasingly focused on the processes which increase a population's susceptibility to disease, since it is these factors rather than exposure to a disease which is critical in explaining spatial variations in health statistics. A study carried out in the 1960s of deaths occurring from measles found that the major contributory cause in 59 per cent of cases was nutritional deficiency. Amongst children and babies, it was particularly evident that

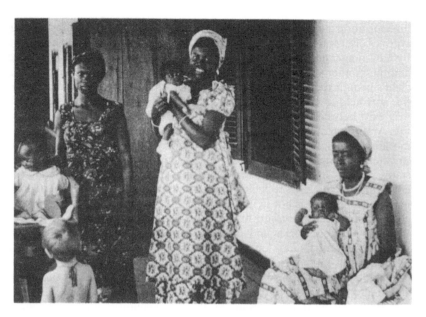

Plates 2.1 and 2.2 Ante-natal, infant and child clinics in Akropong, Ghana
Photos: J Findlay

immunization and medical facilities provided protection only against certain types of diseases, while susceptibility to other diseases is strongly influenced by environmental factors. For example, babies may be poorly nourished and the mother's lack of health education will increase the likelihood of a child's being exposed to health risks. Social and economic determinants of this sort help to explain the vast variations in infant mortality rates which may exist between different districts within Third World cities. A study by the Arab geographer, Kchir (1979), has illustrated variations in infant mortality rates in Tunis from over 160 per thousand in several of the poorest squatter settlements around the city to less than 10 per thousand in the high-status suburbs of the agglomeration. Table 2.1 emphasizes that maternal education is critical in reducing child mortality and that improvement in medical technology is not in itself enough. As the table shows, in Cairo the children of educated mothers have much higher life expectancies because of the very different social and economic environments in which they are reared. Differentials in infant mortality occur not only because of a higher likelihood of immunization against disease amongst children of well-educated mothers but also because of better nutrition, housing, sanitation and greater care in the purchase and preparation of food.

In summary, mortality rates in less developed countries have experienced a substantial drop in recent decades and this can certainly be linked with the beneficial effects of western medical knowledge and health care programmes. Predictions made in the late 1960s and early 1970s that the life expectancy of Third World populations would soon converge with those of European and American populations are now being viewed as over optimistic, since sickness and mortality are in part related to social and economic determinants. Malnutrition remains widespread. In 1981 none of the world's twenty lowest-income countries, with the exception of China, were able to provide enough food to give their populations the minimum

Table 2.1   Cairo: Child mortality and mother's educational attainment, 1976

Educational attainment of mother	Number of children dying before age of two per thousand births
Illiterate	143
Reads and writes	114
Primary education completed	91
Secondary education completed	67
University and above	52
Cairo average	124

Source: Tecke, B. (1985) 'Determinants of child survival' in F. Shorter and H. Zurayk (eds) *Population Factors in Development Planning in the Middle East*, Cairo, Population Council. 137–150.

recommended calorific intake. While amongst the citizens of Third World nations the most educated and privileged members will continue to experience improved life expectancies, the poorer masses are not likely to be able to afford access to the best medical attention. They do not have the incomes to be able to live in environments with low health risks nor to achieve education of the levels necessary to sustain appropriate health care for their children.

Fertility patterns in the Third World

Fertility trends have been the focus of much demographic research, and particular interest in the topic has resulted from the wealth of information produced by the World Fertility Survey which conducted surveys in forty-one developing countries between 1972 and 1984. For the first time, demographers and geographers had reliable and detailed fertility data which permitted international comparisons of fertility levels, family size ideals and knowledge and practice of family planning.

The World Fertility Survey shows that fertility rates in most Third World countries are still extremely high compared with more developed countries. In Asia and Latin America the World Fertility Survey found a total fertility rate of nearly 5 (the average number of children a woman would have if current fertility patterns were to continue during her child-bearing age span), while in Africa it was nearly 7. There is evidence in those countries which participated in two rounds of the survey of a decline in fertility with time. For example, in the Dominican Republic the total fertility rate measured by the World Fertility Survey fell from 5.7 in 1975 to 4.1 in 1983, while in Jamaica it fell from 5.0 in 1975/76 to 3.5 in 1983. The World Fertility Survey showed, however, that fertility decline was far from a universal trend. In Africa, south of the Sahara, fertility rates appear to be still rising in many countries because of improved maternal health and nutrition and a reduction in sterility and breast feeding.

Of considerable interest in the World Fertility Survey is a comparison of the total fertility rate with the desired family size of women of different ages. Figure 2.1 shows the results for some African and Latin American countries. In African countries not only are total fertility rates high, but the average size of family desired is in most instances as high or higher than the total fertility rate. Fertility rates were not only lower in Latin America but the average desired family size by women in the 15–19 age cohort was considerably less than the current total fertility rate, giving rise to the expectation that fertility rates in this part of the world may continue to decline.

It is pertinent to ask which factors are most important in influencing fertility levels and fertility trends. Cultural and economic forces once again operate indirectly through specific demographic variables to affect fertility.

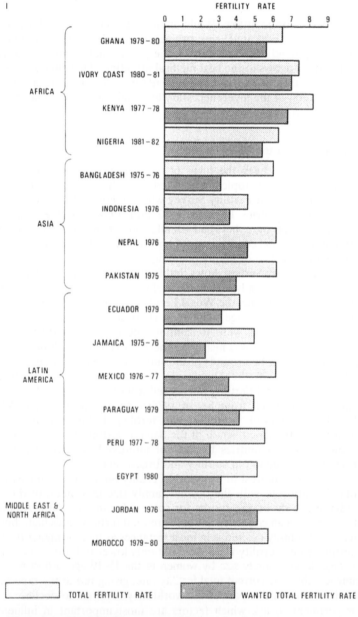

Figure 2.1   Total fertility rates and desired family size
Data source: World Fertility Survey

The demographer Bongaarts (1985) has suggested that temporal and spatial variations in fertility are dominantly influenced by four so-called fertility variables' – the proportion of married females, contraceptive use and effectiveness, the prevalence of induced abortion, and the duration of the period after birth when a woman cannot conceive. The first factor reflects the proportion of the female population likely to have children (the term marriage usually includes consensual unions). Clearly in a country like India, where the age of first marriage is extremely young, averaging 13.2 years at the beginning of the century and rising to only 17.2 years by the 1960s, the potential for large families is extremely high, with the average number of births being greater than in societies where marriage occurs at a later age. Mean age at marriage in the mid 1970s in Bangladesh was even lower than that of India, being only 16.3 years. In addition, the proportion of women never married remains extremely low in countries such as India where only 0.5 per cent of women in the 45–49 age cohort have never been married, compared with 15 to 20 per cent of women in these cohorts in north European countries.

The second and third variables listed by Bongaarts measure the prevalence of deliberate marital fertility control and the fourth is determined by natural marital fertility. Differences between developing countries in the use of contraceptives is striking and is discussed in the final chapter where it is related to the role of family planning in development planning. While spatial and temporal variations at a national level need to be related to cultural and political factors, certain broad generalizations can be made about which elements of a population are most likely to use family planning if it is available. There is considerable consistency in the results of the World Fertility Survey in showing that age, family size and education all influence the likelihood of contraceptive use. It is most common among women in the 30–39 age cohort, particularly those who already have two or three children. Similarly better–educated women are more likely to make use of family planning services, although this relationship is a complex one since family planning services tend to be more available in urban areas where the more educated population lives. In the same way regional variations in contraceptive use are difficult to interpret and, although rural regions have much lower usage, it is important to ask whether this reflects the nature of rural society itself, or is once again a function of the inaccessibility of family planning services. Statistical analysis of the importance of place of residence in influencing fertility does suggest significant differences between rural and urban areas. Fertility levels are different not only because of the better provision of educational facilities in urban areas and because of employment differences between urban and rural labour markets, but are also attributable to specific attitudes to child-rearing found in urban and rural societies.

In any one country the importance of Bongaarts' four intermediate variables inevitably changes through time as societal and economic development occurs. For example, contraception increases in importance through time in reducing the total fertility rate. This fertility 'transition' reflects a switch towards an increased control of fertility by married couples and a reduction of the influence of natural fertility factors.

A lively debate exists amongst researchers as to the social and cultural factors encouraging fertility control in different countries at different times. One of the great problems arising out of the World Fertility Survey is that, while it has permitted considerable progress to be made in the evaluation of intermediate fertility variables, it has not helped greatly in relating demographic trends to patterns of economic development. The timing of a decline in marital fertility is clearly not a direct function of economic development in isolation, but reflects more specifically societal changes and in particular changing attitudes and values towards child-bearing, family size and the role of women and children in the family and in society.

Undoubtedly the introduction of mass education has been one of the most important determinants of the timing of fertility decline in the less developed countries. This is true both because the educational level of the female population is an important determinant of attitudes to fertility control and because the education of children has been critical in altering the social and economic organization of the family in the Third World by transferring children from the role of producers to that of consumers within the household economy.

Case study A

Patterns and determinants of Moroccan fertility

Morocco with 20 million people has the largest population of the north-west African states. The 1982 Moroccan census was the third since the country gained independence in 1956. The 1971–82 intercensal period was marked by an annual rate of increase of 2.6 per cent, a rate very similar to that recorded between 1960 and 1971. Over the last two decades Morocco's population growth rate has presented the country's economic planners with one of their most pressing and intractable problems. Although a national family planning programme has been instituted, its impact remains limited. Consequently the rapid growth of the labour force continues to outstrip the expansion of employment opportunities. With a GNP per capita of $670 in 1984, and a positive rate of economic growth, Morocco has fared better than

Case study A (*continued*)

many other Third World countries in terms of its national economic development. Nonetheless the country has many severe economic problems, not least of which are those relating to population and manpower planning. Spatial variations in and determinants of fertility levels are of considerable importance in the formulation of a development strategy in a country where some 25 per cent of the population is between 5 and 15 years of age and thus will enter the most critical reproductive age cohorts over the next decade, as well as creating a massive new supply of job-seekers entering the labour market.

Fertility rates in Morocco, as in most of North Africa, remain very high. In 1984 the total fertility rate was 4.9. The influence of 'modernization' and of family planning programmes is evident from the findings of the World Fertility Survey where the average desired family size among women in the 15–19 age cohort was only 4.3 children compared with 6.6 among women in the 45–49 age group. The chief influences on fertility in Morocco reflect the effects of age, income and education (figure A.1). Age-specific birth rates peak for the 25–29 age cohort of women in all social groups but, as age at first marriage is considerably less amongst lower-income groups and less educated women, it is not surprising that age-specific birth rates are higher at all ages for low- relative to high-income households. Amongst women with no formal schooling fertility rates are higher than for any other educational group, except for age cohorts of 40 years or more. These variables were also found to be closely related to whether married women had sought family planning advice. Detailed survey work in both rural and urban areas of Morocco has shown that aspects of lifestyle, such as the degree of exposure to the media or the extent of involvement in activities outside the home, have an important influence on fertility levels.

The geographical consequences of these fertility characteristics are evident both in terms of a significant rural–urban differential in fertility rates and in some regional variations. Although high fertility rates do persist among some migrant groups in the cities, rural fertility rates remain considerably higher than those of urban areas. Significant rural–urban differentials exist in the levels of school attendance and illiteracy. For example, illiteracy rates are 82 per cent in rural areas compared with only 64 per cent in urban areas. The very poor north-eastern region, with a very limited physical resource base, has the highest fertility rate in the 15–24 age cohorts as well as in the over-40 age group, while the most industrialized region, which includes the country's largest city, Casablanca, has the lowest

Case study A (*continued*)

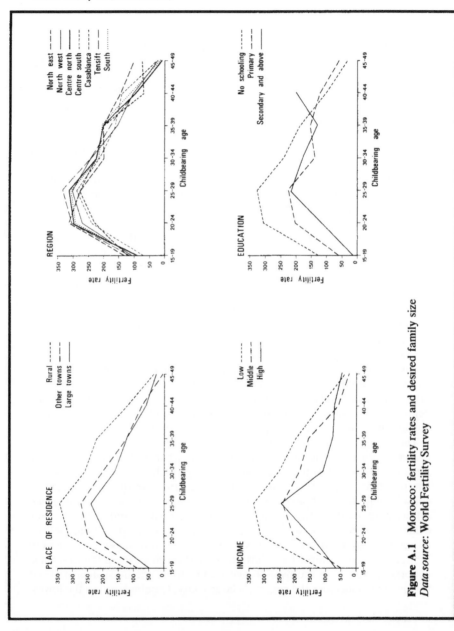

Figure A.1 Morocco: fertility rates and desired family size
Data source: World Fertility Survey

Case study A *(continued)*

fertility rate in the 20–29 age cohort. Within Morocco considerable regional variations in levels of social and economic development exist. As in other Third World countries spatial variations in fertility rates within the country would appear to reflect the indirect influence of social and to a lesser extent economic development on the determinants of intermediate fertility variables. At an international level such generalizations are harder to achieve, with cultural and religious forces being important in differentiating Moroccan fertility patterns from those of other Mediterranean countries.

The findings of the World Fertility Survey can be substantiated by more personal case histories of three Moroccan families. All the families live in the city of Rabat, but none of them is native to the city. Two came originally from rural backgrounds and live in low-income housing, while the other family came from the city of Fez and lives in one of the better parts of the city.

The first family, the Hammoudis, lives in the old Arab quarter of Rabat in a traditional cul-de-sac where they have a courtyard-style house. There are ten people in the family. The father was born in the village of Sidi Zouine in the province of Marrakech. He migrated to Rabat as a teenager, having worked previously as a shepherd boy in his home village. He returned to his area of origin to get married, his wife coming from a neighbouring village. The family tree is shown below, and indicates all the surviving children. Two other children died as infants.

Husband (born 1940)	⎧Son (born 1966) ⎪Daughter (born 1972) ⎪Son (born 1976) ⎨Daughter (born 1978) ⎪Daughter (born 1979) ⎪Son (born 1982) ⎪Daughter (born 1984) ⎩Son (born 1985)
Wife (born 1947)	

Their first son was therefore born when his mother was 19 and, of their eight living children, three were born while she was less than 29. Their father currently works as a self-employed painter, but the eldest son has not yet found work.

The second family, the Tachfines, lives in the same area of the city as the

Case study A (*continued*)

Hammoudis. Once again the head of the household left his village of origin as a young man to seek better opportunities in Rabat. Previously he had lived and worked in a village in the southern province of Agadir, where he had served in his father's grocery shop. His wife came from the same village as himself. They were married and had their first child before migrating to Rabat. The structure of the family is shown below.

Husband (born 1949) Wife (born 1956)	⎧Daughter (born 1973) ⎪Daughter (born 1976) ⎪Daughter (born 1978) ⎨Son (born 1980) ⎪Daughter (born 1982) ⎩Son (born 1984)

By the time the wife was 25, four of the six children had been born. The father currently works as a delivery-van driver in Rabat. His wife returned to their home village for the birth of their second and third children.

By contrast the third family, the Al Fassis, lives in a prestigious part of the former colonial town of Rabat. The father was born in Fez and his wife in the city of Casablanca. He has a white-collar job in the civil service. They have three children as shown below.

Husband (born 1952) Wife (born 1953)	⎧Daughter (born 1978) ⎨Son (born 1981) ⎩Son (born 1984)

As can be seen their first child was not born until the mother was 25 and their children are more widely spaced than those of the lower-income families. This, as well as the smaller family size, conforms with the results of the World Fertility Survey which suggested education and socio-economic status were key determinants of fertility differentials.

It appears from the results of the World Fertility Survey that a small increase in female education in the least developed countries is often associated with a rise rather than a fall in fertility. This occurs because small improvements in education can lead indirectly to improved maternal health, but not to any change in attitudes towards child-bearing. In Third World countries with slightly higher levels of economic development all increases in education appear to cause a reduction in fertility. An inverse association between female education and fertility is one of the most widely substantiated relationships to emerge from the World Fertility Survey. It has many implications for planners and policy-makers interested in the relationship between population and development, since it clearly indicates that social development, in introducing mass education, is a more critical determinant of population trends than is economic development. Since the introduction of mass education programmes is only very loosely associated with levels of economic development, it helps to explain why reductions in fertility levels have occurred in some countries in the absence of any marked advances towards economic development.

Much research has sought to establish more clearly the relationship between female employment and fertility, but generalizations on this topic are difficult to advance. The complexity of the relationship arises because a woman's fertility history itself influences whether she is likely to seek certain types of employment, and it is not possible to suggest that employment affects fertility in a uni-directional fashion. In addition, the work–fertility relationship is influenced by the common determinants of social status and education. A negative work–fertility relationship seems more likely to hold true for certain types of employment. For example, in the Third World, women working in large-scale industrial enterprises and in stable jobs within the urban economy appear to have lower marital fertility than women working in the agriculture or in so-called 'informal' sector jobs in the city.

Relationships between socio-economic variables and fertility remain difficult to establish. One of the most important but problematic issues, highlighted by geographical comparisons between countries, is that socio-economic status appears to be of relative rather than absolute importance. That is to say that, within any specific regional or national context, relative differences in socio-economic status are valuable in explaining why some groups of women have higher or lower fertility levels than others, but, when similar groups of women are compared for different societies or culture regions, the socio-economic variables diminish relative to other constraints and norms influencing attitudes towards child-bearing. At any one point in time it is therefore extremely difficult to explain in socio-economic terms why fertility levels in different countries vary as they do and more sophisticated forms of analysis are required.

Key ideas

1 The decline of mortality rates in Third World countries in recent decades has resulted from improved nutrition, advances in health care and medical provisions and attempts at eradicating many infectious diseases.

2 Maternal education is a particularly important determinant of infant mortality rates.

3 Fertility rates do not appear to be closely associated with rates of economic growth. The most striking feature of Third World fertility rates is their geographical diversity. Social and cultural factors seem to be the key influences on fertility patterns at an international scale.

3
Limited demographic transition

Fertility and mortality are the two major demographic processes involved in national population growth. The causal factors (as opposed to the intermediate variables) responsible for changes in mortality and fertility rates have naturally been the subject of much investigation, as have the mechanisms which link the two processes. The search for causal explanations has often led to attempts to create a 'grand theory' which links together all the elements of population change within what has been termed 'a demographic transition'. This search for a 'grand theory' has been one of the most enduring aspects of population studies and, although it has never met with complete success, the search has proved most valuable in encouraging population geographers and demographers to refine their ideas and to seek generalization from their empirical research.

Notestein's theory

One of the fullest formulations of demographic transition theory was put forward by Frank Notestein in 1945. He based his observations on the changes which had taken place in the fertility and mortality rates of West European countries as a result of the agricultural, industrial and sanitary revolutions which these societies had experienced in the eighteenth and nineteenth centuries. From his observations he concluded that the West European pattern of mortality and fertility decline formed a socio-biological 'model' which might be expected to be found in other countries including those of the Third World. Notestein's theory has been widely studied both from a theoretical point of view and also as a 'model' of the ideal pattern with which specific demographic histories could be compared.

The theory suggested that a transition would take place in mortality and fertility rates from high to low. This latter stage would, however, only be reached following a transitional phase when mortality rates would have declined but fertility rates would remain high. The three main stages in the process are shown in figure 3.1. As mortality rates decline while fertility rates remain high, a period of rapid population growth occurs in the second stage of the model. In the third stage of the model, fertility rates decline because the societal and economic supports to high fertility are undermined as 'modernization' spreads, and as fertility control replaces regimes of natural fertility.

According to Notestein this replacement reflects the increased rationality of fertility behaviour which is associated with so-called 'modern' urban society. Being set within the context of patterns of economic change which characterized the western economies in the eighteenth and nineteenth centuries, the theory in effect interprets the demographic transition as being a response to the processes of 'modernization' which were occurring at the same time. Unfortunately no distinction is made by Notestein between processes of economic growth and those of social change which accompanied urbanization and industrialization, but which had distinctly different impacts on the population characteristics of Western Europe.

As a descriptive generalization of European demographic trends the theory has considerable value and, as will be shown below, is a concept which has been readily transferred to Third World contexts to describe their demographic development. As a predictive and explanatory tool, it is much less satisfactory.

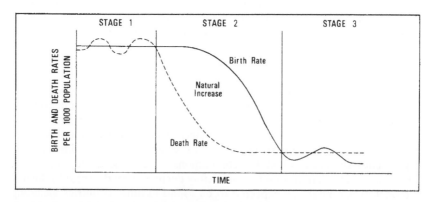

Figure 3.1 Notestein's demographic-transition model

Much time and effort has been spent attempting to classify the demographic regimes of developing countries in terms of Notestein's three stages. The demographic transition, experienced in the past by European countries, is rapidly spreading to the less developed nations. According to Chung (1970), Western Europe, the Soviet Union, North America and Australia had already reached stage 3 of the transition process by 1960. Many Latin American countries were in the latter part of the second stage, while the majority of Third World countries were either still in stage 1 or in the early part of stage 2 of the transition. Although Chung predicted that these countries would all proceed to stage 3 as the demographic transition spread to them, empirical evidence suggests that by the 1980s some developing countries seemed likely to remain in stage 2 much longer than would have been expected.

Instead only a limited demographic transition appears to have occurred (see case study B, pp. 34–8).

Limitations of demographic transition theory

Application of demographic transition theory to Third World countries is difficult for a number of reasons. The most obvious of these is that Notestein's theory was developed from a Eurocentric point of view and it is clear that conditions in Third World countries will never parallel those of Europe in the nineteenth century in terms of either their economic or social development. It follows that, even if demographic changes do reflect trends in 'modernization', Third World countries would differ from those of Europe not only in timing, and speed of development, but also in character, since their economic histories have followed a very different course.

Is there any evidence that the principles outlined by Notestein are applicable? Defendants of the theory point to the widespread evidence in the Third World of declining mortality and fertility rates and emphasize that as yet no country has recorded fertility decline in advance of reductions in mortality. In this respect transition theory has proved robust as a descriptive device. There are also, however, many serious problems in accepting Notestein's theory as an explanatory framework for Third World population change. All the evidence presented in the previous chapter would signal caution with regard to the proposition that rising income levels directly cause demographic change. While death rates can decline as a result of rising standards of living, in Third World countries it would appear often to have been the introduction from outside of medical knowledge and a new sanitary regime which stimulated an initial decline in mortality rates. Similarly the previous chapter suggested that fertility decline was much more closely related to social changes than to economic influences.

A major criticism of Notestein's theory is that it is deterministic, with one stage giving rise to the next in a unilinear fashion without adequate explanation of the mechanisms responsible for powering the transformation. A modification of the demographic transition theory which partly overcomes this criticism was proposed by Kingsley Davis in 1963. He suggested that rather than mortality and fertility rates both responding independently to external stimuli, demographic change is led by a mortality decline and that this in turn stimulates other demographic changes. He proposed that, in order to sustain its standard of living, a population faced with declining death rates and hence rising rates of natural population increase would respond in a number of ways including increased fertility control, as well as rural to urban migration and international emigration. Davis's ideas are interesting since they provide a motor which might help to power the movement of a population through the different stages of the demographic transition, with the causation of demographic trends being internalized, once an initial mortality decline has commenced. Davis's work does not, however, counter some of the other criticisms of Notestein's theory, nor does it explain why and when an initial mortality decline should occur.

Notestein's assumption that societies move towards increased rationality about their fertility behaviour as the demographic transition proceeds, and as superstitions about childbirth and traditional attitudes to family size disappear, has also met with criticism. The Australian demographer Jack Caldwell (1982) has strongly questioned the idea that traditional agrarian societies make irrational decisions about family size. Instead Caldwell suggests that decisions about family size are explained by the nature of the household economy. He proposes that the wealth flows within a household are the critical mechanisms influencing decisions about family size. In traditional agrarian societies, where wealth flows from the children to the parents as a result of their productive contribution to the household economy, it is rational to sustain higher levels of fertility. Caldwell suggests that a fertility transition occurs when wealth flows from parents to children become of greater importance. Consequently as economic and social structures change, and with them the role of the household economy is modified, so a fertility transition can be expected to occur. Caldwell's ideas result from detailed field research in Third World countries, such as Nigeria, and present an interesting perspective on Third World population change, since his work seeks to understand the development processes which give rise to attitudinal change towards fertility. Much work on this complex topic remains to be done and, since it depends on detailed behavioural research, progress is likely to be slow.

A final problem with demographic transition theory is that, like many others, it suggests that population factors respond to development factors.

As later chapters of this book seek to illustrate, demographic variables may have a very significant effect in influencing the course of a country's social and economic development and must be seen as actively influencing a country's development prospects.

Today, the theory of the demographic transition is taken to be an adequate description of the general trend from high mortality and fertility to low mortality and fertility. Although the original theory does not specify properly the causal relationships which exist between mortality and fertility rates and the development process, it remains of value and its suggestion that one day the developing countries will also achieve low fertility is seldom questioned. The transition concept provides a useful starting point for academic inquiry into the 'ideal' sequence which a country's population might be expected to follow. Although it has been shown that fertility decline does not diffuse automatically from the developed to the less developed countries, the very failure of this part of demographic transition theory has helped to promote more detailed research of other mechanisms responsible for fertility decline. The theory of the demographic transition, rather than being rejected on empirical grounds, has tended therefore to be replaced by a number of new sub-theories such as the mobility and fertility transitions.

Fertility transition in the Third World?

Increasingly it has been realized that the demographic process which is least well understood is that of fertility decline, and as the case study of the Islamic world (p. 34–8) indicates it is the cultural diversity of fertility behaviour which accounts for Islamic countries only experiencing a limited demographic transition. Although the demographer Bongaarts, referred to in the previous chapter, has outlined how intermediate fertility variables relate to a theoretical fertility transition in the Third World, the ultimate causes of fertility decline remain poorly understood. The wide-ranging results of the World Fertility Survey present mainly negative findings with regard to any theory of a fertility transition. That is to say, they provide evidence that it is not ultimately caused by the range of economic and demographic variables examined by the survey. For example, the survey suggests that a decline in mortality was certainly not a sufficient condition for a reduction in fertility to occur. This can be illustrated by the cases of Kenya and Indonesia which both experienced similar trends in infant mortality with death rates falling from 16–17 per cent in the first year of life in the 1950s to 9 per cent by the mid 1970s. Despite this common mortality trend, Indonesia had a rapid reduction in fertility levels, while in Kenya fertility rates did not decline and may even have slightly increased during the period under study.

Case study B

Islamic populations in transition

Population data are generally collected on the basis of political units. However, societal, religious and cultural influences, which often strongly affect population trends, may transcend political boundaries as is well illustrated by the case of Islamic populations. Almost a fifth of mankind lives in Islamic countries and Islamic peoples make up the majority of the population in some thirty-seven countries of the world (figure B.1). Inevitably these countries vary in population size and character and indeed in the proportion of the total national population which is Islamic. Equally they span several different culture areas. They do share, however, common

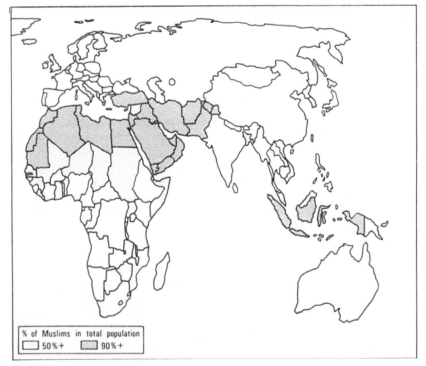

Figure B.1   The Islamic world

Case study B (*continued*)

socio-religious features which have a very significant impact on their demographic characteristics. Many of these countries exhibit the demographic characteristics of other developing nations and, although Islam should not be regarded as an exclusive influence on their demographic development, it does contribute to creating a distinctive demographic environment.

Islam has probably shown more concern for the family than any other religious or social system. Consequently it has a particularly profound impact on societal attitudes towards demographic matters. Islam is not merely a religion but also a way of life. It has laws for nearly all human activities and, at least in theory, does not leave room for a separate secular jurisdiction. This makes it difficult for the government to interfere with, for example, family structures without being accused of religious intervention. With the spread of Islam the family unit replaced the former tribal unit as the dominant social grouping and thus it is not surprising that, as a religion, Islam has developed very specific ideas about the nature of family life, the roles of men and women in society and attitudes towards child-bearing. Islamic countries have continued to favour the early marriage of women, large families and high fertility, and as a result most are characterized by rates of natural increase above those of their Third World counterparts. As the population geographer John Clarke (1985) has shown, demographic transition appears to have taken place to only a limited extent in these countries, even compared with other developing nations. Many of the comparative demographic statistics quoted in the remainder of this case study are taken from Clarke's distinctive contribution to the study of Islamic populations.

Mortality rates in Islamic countries are very varied with crude death rates ranging from 4 to 28 per thousand and infant mortality rates of 31 to 205 per thousand (figure B.2). Death rates are highest in Saharan Africa and parts of southern Asia and lowest in the oil-rich states of the Arabian Gulf such as Kuwait. In Muslim countries it can be expected that mortality rates will decline further with the exception of countries such as Kuwait and the United Arab Emirates where existing access to medical care and good health facilities has already resulted in a very low death rate.

Fertility rates remain by contrast very high with crude birth rates of over 40 per thousand in many of the Islamic countries (figure B.2). There are some notable exceptions, such as the countries where Muslims are less dominant. Turkey, Tunisia and Indonesia are also exceptions and all have strong family planning programmes. They have total fertility rates of 4.0 to

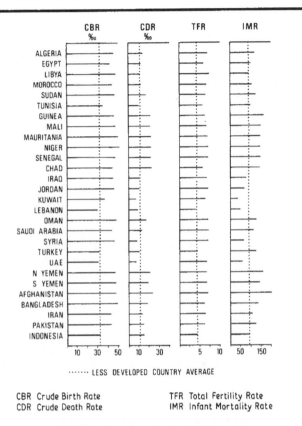

CBR Crude Birth Rate

CDR Crude Death Rate

TFR Total Fertility Rate

IMR Infant Mortality Rate

Figure B.2 Fertility and mortality in selected Muslim countries

5.6 compared with rates of 6.0 or more in most Islamic countries. Nine out of the ten countries in the world with total fertility rates of over 7.0 are Islamic countries and all but three Islamic countries have total fertility rates which are higher than the average for the less developed countries. It should be noted that the Gulf economies, although they have low mortality rates, do not have low total fertility rates. In Kuwait, for example, the total fertility rate in 1983 was 6.1. Figure B.3 shows a comparison of crude birth and death rates for 1960 and 1982. Although crude birth rates have declined in a number of the countries, crude death rates have declined more rapidly.

The proportion of the population aged less than 15 was in most cases over 40 per cent and in a number of countries was 45 per cent or more in 1983. Most Islamic countries will, therefore, continue to experience rapid

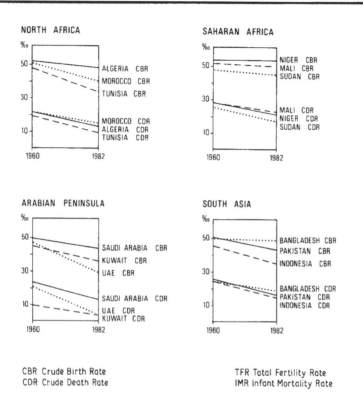

CBR Crude Birth Rate   TFR Total Fertility Rate
CDR Crude Death Rate   IMR Infant Mortality Rate

Figure B.3 Demographic change in selected Muslim countries
Data source: J Clarke, 1985, Islamic populations, *Geography*, 70, 118–28

population growth. Clarke has estimated that for twenty-six of the countries the doubling time is 20 to 28 years. An examination of changing death and birth rates between 1910 and the early 1980s for Egypt has indeed shown a limited degree of demographic transition. Although death rates have declined since 1940, the birth rate has not declined substantially. Infant mortality rates have shown a particularly rapid decline from 250 per thousand in 1932 to 100 in 1982, but the total fertility rate, which did decline in the 1960s from 6.9 to 6.0, appears to have stagnated at 5.6 since 1972.

The cultural context is undoubtedly one of the reasons for the persistence of high fertility rates in Islamic countries. In particular, the role of women in Islamic society may be seen to have a very significant effect on fertility

Case study B (*continued*)

levels. It is difficult to quantify the status of women but a selection of key variables collectively indicate a socio-economic environment which favours high fertility rates and large family size. For example, female literacy rates remain low and are significantly lower than those of the male population. In more than half the countries for which data are available, illiteracy amongst the female population is over 70 per cent with figures as high as 99 per cent in Niger, Somalia, Chad, North Yemen and Afghanistan. The proportion of women enrolled in education in the 12–17 age group was generally below 30 per cent and in several instances less than 10 per cent. Equally female participation in the labour force was in the majority of countries less than 10 per cent, with the exception of West African countries where female participation was significantly higher. In addition, the proportion of the 15–19 age cohort of women who were married displayed a tendency for very early marriage, particularly in countries such as Libya, Chad, Niger, Mali and Bangladesh where it was over 70 per cent.

Although Islam is not opposed to family planning programmes, religious conservatism and the importance of Islamic rather than secular law with respect to family matters have tended to mean that this has not been a high priority. Many governments in Islamic countries perceive fertility rates as too high and thus wish to encourage family planning programmes. The most pro-natalist countries are, however, those of the oil-rich Gulf States where shortages of manpower have created problems in executing development strategies. Increasingly demographic realities have been affecting government attitudes and resulting in the formulation of specific population policies. The success of programmes such as those carried out in Tunisia and Indonesia are evidence that these are compatible with Islamic society, although they have necessitated radical changes in attitudes to the role of women. It should therefore be stressed that the degree of societal change which would be required in an Islamic society for complete demographic transition to occur is considerable. It therefore seems much more likely that the Islamic environment will continue to have a very significant impact on demographic factors and will limit the extent to which demographic transition is likely to occur in Islamic countries, even where demographic pressures are acute. Where such pressures do not exist, as in the oil rich states, it appears that pro-natalist attitudes might typify Islamic society. No group of countries illustrate better than the Arab oil states that rising income levels and higher standards of living do not in themselves result in a fertility transition.

Research points to the conclusion that cultural values and societal norms are all-important in determining fertility trends. Although there is little doubt that the spread of the ideas of fertility control and family planning occurs more quickly as transport and communication become easier in the Third World, and as levels of literacy and education rise, resistance to these changes is geographically diverse because of the greater ability of some cultures and societies to withstand the penetration of western ideas about family size and fertility norms. It is surely this cultural diversity which helps to explain why Islamic countries, for example, continue to have much higher fertility levels than many other developing countries. This conclusion is far from unsatisfactory for the student of development processes, since it points to the very real potential which lies within any given society to chart out its own course of demographic development, rather than to have it dictated by external economic circumstances. Although certain aspects of population change in Third World countries appear unavoidable in the long run, as shown by demographic transition theory, the timing and character of these changes are by no means predetermined. People and the society to which they contribute remain the key factors influencing the course of demographic development.

Key ideas

1 Demographic trends in Third World countries do not appear to have followed the same course as those in developing countries.

2 Attempts to design a 'grand theory' of population change have tended to be replaced by explanations of components of population change.

3 Third World countries have experienced only a limited demographic transition, because of the importance of social and attitudinal forces in sustaining higher than expected levels of fertility.

4
Population and food resources

There can be little doubt that rapid population growth has a significant influence in shaping patterns of economic development in the Third World, but many different views have been expressed as to the precise role of population change in the development process. Many developing countries have managed to sustain increases in their population's average income per capita at the same time as experiencing rapid population growth, thus illustrating that economic and demographic growth are not exclusive processes. It is important, however, to ask whether rapid population growth reduces the pace of development, or whether it has the inverse effect, acting as a stimulant to the implementation of more efficient development policies.

These issues are tackled in this chapter by analysing the relationship between population and agricultural resources, while in the next chapter the population–resource balance is examined in the broader context of population change in conditions of industrialization and urban development. In relating population change to both agricultural and industrial development it is important to realise that technology is of critical importance. The standard of living of a population is not simply a function of the ratio of physical resources to population size but is much more fundamentally related to the stock of information and technical ability which is available to a population to make agricultural and industrial raw materials of some direct value to society. It is only then that these physical materials can truly be termed 'resources'. For example, in the last thirty years the volume of grain produced by the world's agricultural systems has doubled. This does not reflect any significant change in the environmental or soil resources of the earth's surface, but rather it illustrates the improvements in technology which have permitted us to unlock much greater potential resources from agricultural systems.

Concern over the depletion of the world's scarce resources is not new and many have forecasted long-term problems in feeding the world's expanding population. One of the most famous writers on the topic was Thomas Robert Malthus who published his now famous *Summary View of the Principle of Population* in 1830. The basis of Malthus' theory was that population, when unchecked, would increase geometrically while the means of supporting this population would increase only in an arithmetic fashion. Consequently he foresaw the emergence of a chronic imbalance between population and physical resources, with population growth rapidly outpacing the means of supporting even a subsistence lifestyle. The consequences of such an imbalance were, in Malthus' opinion, likely to result in vice, famine or war. One or more of these evils would occur where an intolerable mismatch of population and the means of subsistence was found and as a result mortality rates would rise until population numbers were once again in equilibrium with resources. Although Malthus himself later retracted some of his more pessimistic views of the future of the human race, he became best known for his belief that 'overpopulation' could only be overcome in these ways, and that there was therefore a limit to the extent to which population growth was possible.

Limits to growth?

The ability of the earth's physical environment to support mankind varies in time and space. This is true, on the one hand, because of geographical variations in the distribution of physical resources and, on the other hand, because of differences in the availability of appropriate technology to utilize resources and in the way that human beings choose to adapt to their environment in the form of different types of social organization. Given any specific level of knowledge and form of social organization, there are limits to the numbers of people who can be supported directly by the resources available within a limited geographical region. These limitations were illustrated in an extreme form in 1984 and 1985 in the terrible famine experienced by the populations of the Sahel countries such as Ethiopia. The development of the Ethiopian famine disaster with tens of thousands dying of starvation might appear to be a classic example of a Malthusian situation where population exceeded available food resources and famine ensued (plate 4.1). Although Ethiopia is a land of moderately rich agricultural potential, throughout the 1970s and early 1980s certain localities of the country suffered persistent drought conditions. The worst affected were the highland regions of Wollo, Gondar, Shoa and Tigray where some 20 to 30

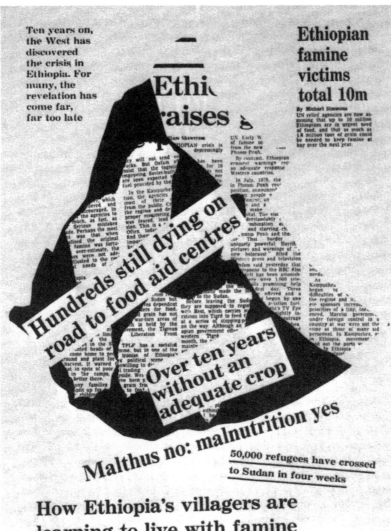

Plate 4.1 The Ethiopian famine: in what sense a 'population' crisis?

million Ethiopians live, and where the majority of the population has been threatened with total starvation. There is, of course, nothing new about drought in Ethiopia. It is a country which has always been periodically afflicted by drought, but the scale of the problem in the 1980s was greater than ever before, because of the larger size of Ethiopia's population, the extraordinary length of the drought and the regrettable use made of human and physical resources in the country. Evidence of this third problem is illustrated by the fact that food production per head of population actually dropped by about 5 per cent per head over the six years preceding the drought. A Malthusian interpretation might appear to have support if the circumstances of the immediate locality were investigated in isolation, but Malthus' deterministic stance is not justified if the Ethiopian situation is viewed within a global environment. Population growth and regional development cannot be isolated from the national and international contexts in which they are set.

The Ethiopian famine led to massive refugee movements of people whose home villages could no longer support them. Similar refugee migration from the war zones of the world has avoided the worst effects of some of the other so-called Malthusian 'checks' to population growth. Population mobility is only one solution to regional imbalances in population/resource ratios. A more satisfactory solution in the short term is the movement of food resources into areas which are food deficient and in the long term the exchange and trading of equipment, ideas and technology to permit the increased agricultural potential of these regions.

While drought may be unavoidable, famine is totally avoidable. Research shows that the famine in Ethiopia, like famine in most parts of the world, was not a sudden event but the result of the successive failure of many harvests and that the problems of the area had been accumulating for many years prior to the crises of 1984 and 1985. Food and agricultural experts have shown that the world production of cereals and other key food items is more than capable of matching the world's food needs. It has been inadequate investment in certain Third World countries in agricultural development, combined with failures in communication and inefficiences in the way in which food resources have been redistributed, that have been mainly responsible for the persistence of famine in the 1980s.

Fortunately most countries are close to being self-sufficient in terms of food (figure 4.1). Some Third World countries have, however, found that they have had inadequate capital to invest in rural development projects and have chosen instead to use their capital resources in other ways. In the case of Ethiopia a long civil war has drained the country of resources which could otherwise have been invested in rural infrastructure and agricultural development.

On a national level continued rapid population growth and the persistence of

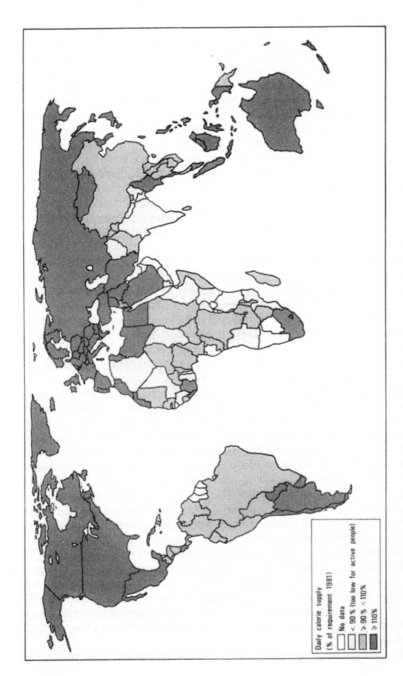

Figure 4.1 Unfair shares? Daily calorie supply as a percentage of requirement 1981

inadequate agricultural development have led to chronic food shortages in no less than 27 of the 39 countries of sub-Saharan Africa. For many of these countries, like Ethiopia, the only way to avoid major food crises is to seek external aid in order to pay for food imports and to finance development projects which offer a long-term solution. Dependence of this sort raises many serious national and international political and economic questions and points to the need for the more developed countries to accept some responsibility for helping to solve the long-term development problems of the least developed countries.

It has been estimated that by 1990 the countries with a food deficit will require the net import of about 100 million tonnes of grain. This figure represents requirements which would meet expected demand rather than the amount needed to ensure a reduction in undernourishment. The international Food and Agricultural Organization (FAO) suggests that this demand for grain by the importing countries can technically be met by the exporting countries as long as the political will exists to achieve this, and that some reserve capacity in fact exists in countries such as the United States which could further increase production. Whether importing countries can, or choose to, afford to import the grain they require remains a major issue. Where nations appear to have no choice but to rely on aid rather than trade to meet their food needs, the transfer of other resources and technology to achieve long-term improvements in their agricultural economy becomes an even more complex issue.

The Ethiopian famine once again exemplifies these difficulties. In 1982 the Ethiopian government made its first requests for food aid. Partly because of the country's political history and partly because of communication problems, there was minimal response at an official level to these early requests for assistance. Even in 1984 and 1985 much of the food aid channelled to the area came through the initiatives taken by non-governmental and charity organizations rather than from the western governments who controlled access to most of the world's food surpluses. In summary, the disaster in Ethiopia and Sudan, and indeed in other Sahel countries, arose not on account of Malthusian factors but rather due to failures in communication and the lack of a national and international political will to reduce the severe imbalance between population and food resources.

Although Malthus' thesis is inadequate because it ignores the roles of technology, trade and population mobility, it remains ultimately true that population growth necessitates increased food production. Since increased food production only occurs as a result of increased investment in agricultural production, it appears to be the failure to achieve this investment, combined with the inability of many poorer countries to be able to organize and pay for food imports, that has led to Third World populations being characterized as suffering from malnutrition and under-nourishment. In many of the countries experiencing very rapid population growth there has been an inadequate increase in food production to meet requirements, while the countries which export grain have on the whole been the ones with much lower rates of population growth. The FAO proposals on how the world can feed itself therefore necessitate both an increase in the use of technology and a change in trade patterns to ensure that food resources are available. to the populations who require them. It is not easy, particularly in view of the experience of the Ethiopian famine, to see how these changes will come about during the next few decades. There are, however, some encouraging examples, such as the case of India, which show that countries with rapidly expanding populations can make remark-able advances in increasing their agricultural production and in reducing the worst effects of drought on their population.

Case study C

India: Feeding the world's second largest nation
by A. J. Jowett

India provides an interesting case study in balancing the equation between 'people production' and 'food production'. At the turn of the century India had a population of 238 million and at that time high death rates cancelled out the high birth rates and thereby maintained a relatively stable population (table C.1). Slight increases in total population accrued during periods when the forces maintaining high death rates temporarily abated, as was the case between 1901 and 1911. However, these increases in population were soon checked by famine and disease which so increased mortality levels in the decade 1911/21 that India's population went into decline. During the decade the *subcontinent* suffered several famines but the major disaster is generally considered to have been the influenza epidemic in 1918/19 which was responsible for an estimated 18 million deaths. The severity of the epidemic was heightened by the drought of 1918 which reduced foodgrain production by over 30 per cent and precipitated a 50 per cent rise in food prices in the following year. It must be remembered that people in less developed countries, weakened by prolonged under-nutrition, are unable to withstand what in the developed world are but minor ailments. Furthermore, even in normal years, there are large sections of India's population who cannot afford an adequate diet. At times of food

Case study C (*continued*)

Table C.1 India:[1] Demographic characteristics, 1891–1981

	Total population (millions)	Birth rate[2] per thousand population	Death rate[2] per thousand population	Infant mortality per thousand live births	Growth rate per year (%)
1891	235.9	48.9	41.3	–	–
1901	238.4	45.8	44.4	–	0.01
1911	252.1	49.2	42.6	287	0.56
1921	251.3	48.1	48.6	290	0.03
1931	279.0	46.4	36.3	241	1.04
1941	318.7	45.2	31.2	211	1.33
1951	361.1	39.9	27.4	n.d.[3]	1.25
1961	439.2	41.7	22.8	183	1.96
1971	548.2	41.1	18.9	146	2.20[4]
1981	685.2	37.0	15.6	122	2.25[4]

Notes: 1 The partition of India and Pakistan in 1947 results in a discontinuity in the statistical series.
2 Data refer to intercensal decades, e.g. 1891 = 1882/91.
3 No data.
4 Corrected for the timing of the census.

scarcity and inflated food prices, these people suffer disastrously. The death rate among low-caste Hindus in Bombay in 1919 was 218 per thousand.

Since 1921 the gradual control of epidemics through the introduction of improved medical standards, improved nutrition and improvements in national and international transport networks, have contributed to a sustained decline in mortality, especially infant mortality. This decline in mortality triggered off a rapid increase in the growth rate of population and in the post-independence era India's population has more than doubled, from around 350 million in 1947 to more than 750 million in 1986. During this period population growth was not halted by famine as Malthus' thesis would suggest. Indeed India has suffered only two minor famines in the past half century, the Bengal famine of 1943 and the Behar famine of 1965/66. In the former case wartime conditions greatly aggravated the problem of organizing and distributing relief supplies, but in the mid-1960s massive shipments of grain from the United States, allied to India's impressive capacity for operating relief programmes, averted a major disaster.

Contrary to Malthusian claims and the popular view, India has managed

Case study C (*continued*)

to increase foodgrain output at approximately the same rate as population over the past thirty-five years (figure C.1). In the 1950s and early 1960s the growth of foodgrain production was achieved by expanding the area devoted to foodcrops and by achieving a better use of traditional technology. This period was brought to an end by the severe droughts of 1965 and 1966. Food production declined by almost 20 per cent and grain output in 1965, at 72.3 million tons, was exactly equal to that of 1953, but in 1965 the population was 110 million greater than in 1953. To prevent widespread starvation India had to import 19 million tons of grain in 1966 and 1967. This period highlighted the vulnerability of Indian agriculture to adverse weather conditions and raised doubts over the wisdom of government policy which in the 1950s and 1960s aimed to offset any shortfall in annual production by food imports rather than pursuing the more costly policy of storing large reserves of grain.

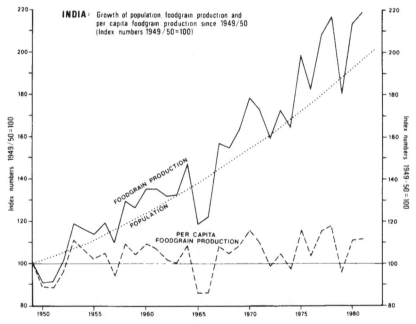

Figure C.1 India: grain production and population growth

Case study C (*continued*)

In the wake of the droughts of 1965/66, with rising imports and threatening famine, a new strategy of agriculture production was devised with the objective of accelerating foodgrain production and achieving self-sufficiency. The new technology, the so-called 'Green Revolution', based on high-yielding varieties (HYV) and an increased application of irrigation water and chemical fertilizer, proved an instant success, especially so in the case of wheat. The area planted to HYV of wheat increased from 0.5 million hectares in 1966/67 to 10.0 million hectares in 1972/73 and wheat production rose from 11.4 to 26.4 million tons between 1966/67 and 1971/72. These developments were particularly marked in the north-west of India, especially in the states of Punjab and Haryana.

Table C.2 Impact of drought on foodgrain production 1918, 1965 and 1972

	1918	*1965*	*1972*
All India average annual rainfall (cm)	85.6	86.9	84.9
Departure from normal rainfall (%)	−21.6	−20.3	−22.2
Area of India under deficient rainfall (%)	60.1	52.0	63.5
Impact on foodgrains			
Change in production (%)	−32.3	−19.6	−8.2
Change in grain prices (%)	53.9	12.2	18.0
Population (millions)	251.4	482.5	563.5

Source: Mooley, D. A. *et al* (1981).

The 'Green Revolution' of the late 1960s, while it generated significant increases in production, also resulted in the widening of interregional and interpersonal disparities which, it was feared, might ignite social unrest in rural India. Thus in the 1970s the government moved to a policy of growth with social justice, which aimed not only to maximize production but also to enable as large a section as possible of the rural population to participate in the benefits of development. In addition the Indian government switched from a policy of food imports to the development of domestic bufferstocks in which surplus grain produced in years of bumper harvests was stored to cover shortfalls in times of drought. Some indication of the success of these policies in the 1970s can be seen in the drought of 1972; this was more severe than that of 1965 and fractionally worse than that of 1918, but the downturn in production was only 8 per cent (table C.2). Similarly in the drought of

Case study C (*continued*)

1979, when production declined by over 20 million tons, there was no return to the famine conditions of 1965/66, for on this occasion the country was protected from disaster by the large food stocks held by the government. In fact, the outside world seemed totally unaware of the magnitude of India's droughts in 1972 and 1979. By the end of the 1970s India was claiming self-sufficiency in foodgrain production.

The evidence of figure C.1 indicates that rapid population growth can occur without experiencing the 'checks' predicted by Malthus. Famine can be avoided when technological development, helping to increase foodgrain production, is accompanied by improvements in food storage, marketing and redistribution. In the early 1980s India was in the embarrassing position of experiencing problems in storing its surplus grain and by the mid-1980s India was even providing a small quantity of grain to famine-stricken Ethiopia, thereby reversing its role from food recipient to food donor.

While acknowledging the significant progress made in increasing food production in India since independence one should not lose sight of the many problems which remain to be overcome. Two examples will suffice. Firstly, while many experts believe that India has the capacity to produce more than enough grain to feed the 1000 million people expected by the end of the century, there is little doubt that rising population pressure is having a negative impact on the quality of the environment. Expanding the area of cultivated land and instigating a more intensive use of existing agricultural land are aggravating environmental problems through deforestation, soil erosion and the encroachment of arable agriculture into already limited grazing areas. The Indian government estimates that over 60 per cent of the country's arable land suffers from environmental degradation and along the Himalayan arc the tree cover has all but disappeared below 2000 metres.

Secondly, it must be stressed that, when India claims self-sufficiency in food production, she is claiming to have satisfied the economic demand, but not the nutritional demand, for food. Nearly half India's population exist below the poverty line, having an income far too low to be able to afford the recommended nutritional requirements of 2400 calories per person per day in rural areas and 2100 calories in urban areas. In 1980/82 food availability in India averaged only 2030 calories per person per day which represents only a marginal improvement over the situation in 1971 and still falls well below the basic requirement. In India, as in the world at large, hunger and starvation will be overcome only when poverty itself is eliminated, for malnutrition is a poverty problem rather than a population problem.

Boserup's hypothesis

Boserup, a Danish economist, has postulated (1965) a more optimistic hypothesis concerning the relationship between population and physical resources. She has suggested that population growth, rather than being a hindrance to economic growth, is actually a prerequisite for agricultural development. Boserup proposes that extensive forms of land use such as slash/burn agriculture provide more efficient means for supporting low population densities than other agriculture systems when these systems are compared in terms of their labour input. At higher population densities more labour-intensive systems are adopted only because these offer higher total levels of food production rather than higher returns to the individual agriculturalists involved. Her theory can be summed up in the phrase 'necessity is the mother of invention', with the adoption of more labour-intensive techniques and more technologically sophisticated agricultural systems only occurring when forced upon populations by rising population pressures. The logical conclusion arising from Boserup's hypothesis is that population growth leads to development rather than hindering it.

Like Malthus, Boserup assumes that a closed community exists, and it has proved very difficult to test her ideas simply because out-migration so often occurs from population pressure in certain regions rather than from other mechanisms' coming into operation in the way in which she predicted to modify population–resource ratios. As with all theories of this kind it is possible to find examples which conform to Boserup's expectations. Boserup herself chooses examples mainly from tribal societies in equatorial countries which display agricultural innovation along with slowly rising population densities. Her evidence is circumstantial rather than conclusive and it is equally easy to find examples to the contrary. She herself admits that over-rapid population growth can lead to inappropriate methods of increasing agricultural productivity being adopted and can result in land degradation. It is clear that ecological constraints limit the extent to which agricultural development is possible over short time spans in certain types of fragile environment such as the Sahel. It must also be remembered that agricultural adaptation cannot be effected instantly and that successful innovative change takes time to introduce. Meanwhile the rate of population increase and population pressure may overwhelm the agricultural system and prevent positive changes being effected.

Boserup also admits that there are many circumstances which may prevent agricultural development, despite rapid population growth having taken place. For example, she suggests that in India during the colonial period many farmers had no incentive to improve their production because of the feudal land-tenure system which operated in many regions and which meant that farmers did not control their own land. Even when population pressure on land is extreme, farmers may fail to make the appropriate investments in improved agricultural techniques if they feel that it is others rather than themselves who will benefit. This illustrates that not only are the influences of economic development indirect and complex, as shown in earlier chapters, but that the inverse is also true, and population growth, with its increasing demand for food, has complex and indirect influences on patterns of agricultural economic development.

With the development of a world economy, agricultural production has increasingly been organized on a large scale, with regional specialization of production permitting agriculturalists to gain higher value for their crops through exchanging them in a market system rather than trying to grow crops only to meet their own food needs or those of their local community. It has therefore increasingly been market forces which have stimulated changes in agricultural systems, encouraging farmers to cultivate higher value crops. Ironically it is at the level of national governments that Boserup's ideas might come closest to being applicable, since Third World governments, faced with large food import bills for their growing populations, have often become motivated to encourage changes in their agricultural systems. Sadly these developments have often been in the direction of promoting cash-crop production for export to help pay for their food imports, rather than in the direction of improving domestic food production.

Population growth and the Green Revolution

As has been shown, neither Malthus' nor Boserup's hypotheses adequately evaluate the relationships between population growth and development; in particular, both ignore the potential of trade in goods, ideas and technology. Population growth has frequently been accelerated by the transfer of western medical knowledge, but the pattern of demographic transition experienced in the developed countries has been shown to have been only partially transferred to the Third World. If the world is a 'global village', as is often claimed, it is to be hoped that international flows of goods, ideas and technology will take place, assisting in the alleviation of 'overpopulation' and population–resource imbalances.

The nature of the impact of population growth on development varies depending on how societies and political units respond to the challenge presented by the need to increase the supply of food. Some countries have tackled the issue by seeking to develop their mineral resources and industrial production in order to pay for increased food imports, while others have resource policies which have sought to increase the efficiency of agricultural production.

A common route to achieving rapid increases in agricultural productivity has been to introduce foreign expertise and western-based agricultural techniques and particularly to encourage the use of new high-yielding varieties of grain and rice. This approach has resulted in what has been termed the Green Revolution, and has led to some Third World countries dramatically increasing their cereal production at rates in excess of those of population growth. The advantages of introducing these high-yielding grains has been that they are generally more responsive to fertilizers, they have a shorter growing time, yields are much higher and they are less sensitive to wind damage and to the number of hours of daylight. In India production doubled between the mid 1960s and 1970s. However, the impact of the Green Revolution is varied and its potential constrained. It must, therefore, not be regarded as a universal panacea but it certainly has in some instances resulted in substantial increases in the food-production capacity of developing countries and exemplifies the important roles of national resource policy and intermediate factors in the population–resource balance. In India technological advances have made the fastest progress in the environmentally favoured areas such as the Indo-Gangetic Plain. It is undoubtedly true that the larger agricultural units have benefited most as they have been able to afford new technologies and have also had better access to the necessary infrastructure. The Green Revolution has only had a limited impact in some other environments because the new varieties are more sensitive to drought and floods, thus necessitating both adequate water- and drainage-control systems and water supplies. Equally the required fertilizers are often too expensive for small farmers. Ecologists have viewed the new varieties with scepticism because of their lack of genetic variety and susceptibility to pests and disease. Similarly some of the social and economic consequences in terms of employment and organization of agriculture may be regarded as undesirable, although they may also be symptomatic of necessary changes in the organization of society and the evaluation and use of human resources. Some of the main problems are, however, the costs involved which many of the poorer countries cannot afford, particularly in terms of the infrastructure required to take advantage of the benefits of the Green Revolution. The fact that the Green Revolution was the inspiration of the developed world rather than indigenous to the Third World has perhaps been the greatest problem. It has offered a technological solution to increasing food production, the benefits of which have been fully achieved only where planners and development agencies have taken time and effort to adapt external ideas to specific geographical and cultural contexts.

Key ideas

1 The limits to population growth predicted by Malthus do not appear to have 'checked' Third World population increase in the twentieth century.
2 The idea that population growth is a stimulant to innovation and development seems plausible. It is in practice very hard to test.
3 Technological advances have helped Third World countries greatly to increase their capacity for food production, but this has in itself not eradicated problems of starvation and malnutrition.

5
People making a living

Inherent in the development process are far-reaching changes in the way in which people earn a living, and in the types of environment in which different economic activities occur. The total number of people employed in world agriculture was estimated to be 707 million in 1950, rising to 790 million in 1975. This increase was very small relative to world population growth and the relative importance of agriculture in terms of jobs declined during this period from 64 per cent to 48 per cent of total employment. On a world scale this trend reflected the consequences of certain economic changes, the global effects of which were to reduce the proportion of the world's population earning a living from agriculture and to increase the proportions depending on industrial and service activities. Since these latter economic activities are spatially concentrated in specific locations, the consequences of these structural economic changes have been the massive geographical redistribution of population towards certain urban regions, and especially towards the capital cities of the Third World. Of the different parts of the Third World, Latin America had the highest levels of urbanization (61 per cent) in 1975, whilst South Asia and Africa had levels below 25 per cent in 1975. By the year 2000, if current urban growth rates are sustained by continued immigration, there will be 25 cities in the world with more than 10 million inhabitants, and of these 20 will be in developing countries. Urbanization rates are expected to accelerate fastest in Africa where by the year 2000 some 42 per cent of the population will live in urban areas. In Latin America the proportion of the population living in urban areas by 2000 will be very similar to that of the more developed regions.

The following chapter examines the demographic aspects of this enormous change; other books in this series, by David Drakakis-Smith and Graeme Hugo, look more closely at the nature and impact of the movement on rural and urban areas themselves.

Rural–urban migration in the Third World

The patterns of population movement and the ways in which population redistribution has occurred in the less developed countries have been strongly influenced by factors external to the economies and societies of the Third World. For example, the stimulation of much large-scale rural–urban migration can be traced to colonial contacts. European imperialist powers viewed Third World countries as a source of raw materials and as an outlet for selling western manufactures. As a result the spatial structures of most Third World economies became strongly focused on a small number of port cities. It was on these cities that newly established transport systems concentrated and it was towards these nodes that rural–urban migration occurred. Rural–urban migration did not, therefore, necessarily occur to all urban areas in Third World countries, but specifically to those centres where external economic influences acted as a catalyst to a western form of 'modernization'.

With the end of the colonial era population redistribution towards these centres did not cease; indeed, it increased. This once again has been due to the nature of economic interaction between the developed and less developed economies. From about 1960 onwards there has been an increased tendency for European, American and Japanese companies to open factories in Third World countries. This has not only been to serve local markets, but to export manufactures to the older industrial nations. This transfer of industrial production to the Third World has happened for many reasons, but one important factor has been the much cheaper cost of labour in the less developed countries. Multinational companies have been particularly eager to employ cheap female workers in industries such as textiles, which are labour intensive. Inevitably most of the branch plants of these foreign companies have once again located in coastal zones close to large urban agglommerations, thus adding to the existing spatial inequalities inherited from the colonial era. The consequence, in terms of population and development, has been a strengthening of the attraction of these more 'developed' urban regions to migrants from the less accessible and poorer rural regions.

Mabogunje (1970), a Nigerian geographer, has produced a migration theory which attempts to illustrate the mechanisms influencing rural–urban migration. He has described rural–urban population transfer as a 'circular, interdependent, progressively complex and self-modifying system in which the effects of changes in one part of the system can be traced through the whole system'. Mabogunje suggests that migration reflects the complex and changing interaction of a variety of forces, such as the individual human personality, as well as the influences of the social, physical and technological environment. The perception of opportunities to earn a higher wage, or to achieve a more desirable lifestyle in urban areas, is likely to attract migrants from surrounding rural areas, and these migrants themselves become sources of information about urban opportunities for their families and neighbours living in rural areas. If the information passed from urban to rural areas by migrants is positive, this is likely to generate further rural–urban migration. In addition to information conveyed back by family members, friends and associates, other evidence of a migrant's success in the city may be transferred back to the area of origin.

The role of remittance money too has been shown to be of vital importance. Urban–rural transfers of money by migrants can encourage further migration both directly and indirectly. In a direct way, by sending money to their villages of origin, migrants demonstrate the success of their new urban lifestyle and thus encourage their families and friends to join them. Indirectly, one of the most common uses of migrant earnings is to help pay for the education of other members of the family, a process which leads to a changed appraisal of the opportunities available. The way in which migrant remittances affect societal and regional income distributions remains poorly understood, but it is quite possible that there may be a connection between migrant remittance networks and the societal transition to family nucleation. The above mechanisms are just some of the ways in which Mabogunje suggests that the migration process is self-reinforcing and thus results in the further redistribution of population from rural to urban areas. Economic changes may initially stimulate migration, but in the long run population redistribution itself results in regional economic changes.

The selective impact of migration

Mabogunje's model is particularly useful in understanding the nature of interactions in the rural–urban migration process. However, the impact of the rural–urban migration process varies considerably according to the nature of migration, the scale at which migration is occurring and the ability of the urban area to absorb the rural influx. Increasingly in the Third World, migration to large cities is occurring not just from rural areas but also from smaller cities. The impact of rural–urban and urban–urban migration has important demographic and regional development consequences, as case study D (pp. 58–61), about Tunisia, shows.

The migration process is naturally selective, attracting certain sections of the population more than others. Consequently it has very specific

Case study D

Tunisians on the move

Almost all the people you meet in Tunis appear to be migrants, all with their own personal migration history of how they came to the capital city in search of work. I first met Ahmed when I was staying in a 'hotel' on the edge of the old city of Tunis. The hotel was in fact no more than a dormitory with eight to ten people to a room. It was occupied by young men who had come to Tunis to find work and who either had not yet found a stable job or could not afford to rent other accommodation. Ahmed was 18 and was working as the hotel receptionist. This was his second job in the capital and he had been working as a receptionist for several months. When he had first arrived in the city, he had stayed with his uncle, also a migrant from the same village in southern Tunisia. His first job had been selling food from a street trolley, but he frequently found that he made insufficient money from this to even cover the cost of his own food. Although the job in the hotel gave him minimal wages, far below the national legal level for Tunisia, at least it assured him of somewhere to sleep and the hotel kitchen provided his meals. He hoped that one day he might get a better job in one of the tourist hotels and receive tips from European visitors.

It was several months later, when I was travelling in southern Tunisia, that I chanced once again upon Ahmed. He had a bandage around his arm and explained that he had lost the job in Tunis after an accident in which he had broken his arm. As soon as it was better, he would go back to the city and stay again with his uncle, providing his parents would give him the money for the trip. This seemed fairly likely since there was nothing for him to do in his home village, and his parents were supported mainly by the remittances of Ahmed's older brother who now lived and worked in France and was considered by the whole family to be very successful, since he had obtained a job on the assembly line of a Renault factory near Paris.

Personal histories such as these epitomize the nature of rural–urban migration. Most rural migrants depend on financial and social support provided by their families and kin during the early phases of urban employment and far from all migrants succeed in integrating themselves in the urban economy.

Figure D.1 shows the pattern of interregional migration which occurred in Tunisia in the years 1969 to 1975. The rural–urban migration process described by Mabogunje is evident, with the largest migration flows focusing on Tunis from the rural north-western provinces of the country. The volume of in-migration to Tunis declines as would be expected with

Case study D (*continued*)

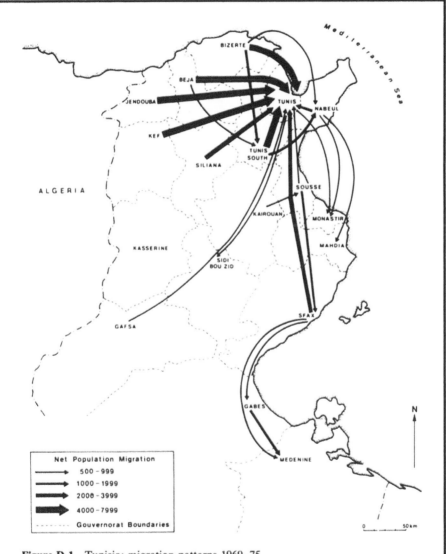

Figure D.1   Tunisia: migration patterns 1969–75
Source: A Findlay, 1982, Migration Planning, *Applied Geography*, 2, 224
(reproduced by permission)

Case study D (*continued*)

Plate D.1   Tunis: migrant housing in the squatter settlement of Saida Manouba

distance. A second dimension of Tunisian migration patterns which is of interest is the increasing proportion of movements which are occurring from small and middle-sized cities to the major urban centres. These urban–urban flows accounted for nearly half of all Tunisian migration movements in the 1975 census compared with only a quarter of the moves recorded in the previous Tunisian census. This emphasizes the fact that migration is towards the larger more 'westernized' cities where the effects of economic growth have been concentrated, and shows that urban areas in themselves are not necessarily attractive to migrants.

Tunisian migration patterns like most redistributive flows illustrate the expected characteristics of migrant age and sex selectivity: 11.6 per cent of all men between the ages of 25 and 29 were recorded as changing their region of residence compared with only 7.1 per cent of women of the same age. Mobility of both men and women was lower in other age cohorts, with

Case study D (*continued*)

for example only 6.5 per cent of men and 3.6 per cent of women in the 35-to-39 age cohort being migrants.

By 1980 continued in-migration to the Tunis conurbation had brought its population to over a million persons, making the city four times larger than Tunisia's second city. Compared with many other Third World cities, Tunis's growth has been modest and it would undoubtedly have grown even more rapidly, had it not been for the diversion during the 1960s and 1970s of some labour migrants to foreign labour markets, first in France and later in Libya. Although the departure of most of the colonial population in the late 1950s created some job opportunities in Tunisia's four main ports, by the 1960s the urban economies of these settlements had once again become stagnant. It has been estimated that as many as half of all in-migrants to the city are forced to work in small-scale activities such as ambulant traders, street porters or hawkers when they first arrive in the city. The problem of rising unemployment amongst migrants led the Tunisian government to take a variety of measures to discourage further in-migration, including futile attempts at transporting unemployed persons back to their villages of origin. More positive measures, such as the promotion of labour-intensive industrial development through the encouragement of foreign investment in export industries, have also met with only limited success. This has been the case firstly because these types of policies resulted in the creation of jobs mainly for the female workforce and secondly because foreign investors favoured factory locations close to or in the major cities and have consequently increased rather than reduced the country's spatial and economic imbalance.

development impacts on both areas of origin and destination, by redistributing the economically most productive and demographically most fertile elements of the population. This operates to the cumulative detriment of regions of out-migration. Many rural studies have shown that areas with substantial rural out-migration experience a reduction in their productive capacity because of having insufficient labour at harvest time.

It is equally true that areas of in-migration do not 'benefit' in a clear cut way from migration. Not all migrants find urban life to be better than their standard of living prior to out-migration from a rural area. In fact, during the first few weeks and months of living in a city, many migrants find that they have to depend on support from their villages of origin or from other relatives or friends who have already been integrated into the urban economy. A common strategy amongst migrants arriving in the city is to try and sell village produce. A study by the Australian geographer, Graeme Hugo, of rural–urban migration in Java found that the process was responsible for the flow of rural savings to the cities to support jobless and underpaid migrants, as well as the better-known counterflow of savings from the city to the countryside. Migrants were also responsible for bringing about economic change in the rural environment by marketing agricultural produce and, as a result, changing the character of the rural economies from which they came by encouraging the production of crops for sale in the city rather than the production of food needed in the villages.

One of the greatest problems facing governments in Third World countries is the provision of sufficient employment to satisfy the rapidly growing labour forces of the urban areas. In most Third World economies governments seem to have discovered only limited opportunities to absorb surplus labour in rural areas, with most agricultural development projects seeking economic efficiency in food production over goals of job creation. It has been to the urban economy that the under-employed, as well as manpower planners, have looked to find new job opportunities. Ironically this has of course strengthened rural–urban migration in the way that Mabogunje outlined, with continued investment in urban economies bringing forth further internal and international migration.

Some migrants never find full employment in the city but most consider the opportunity of urban living to be preferable to the harsh conditions from which they migrated. Not surprisingly the so-called 'informal' sector of self-employed street vendors and petty traders has expanded rapidly in these cities (see Drakakis-Smith in this series). A city such as Bombay has an estimated 200,000 such street vendors eking out a meagre existence. The development problems of a large migrant population are not restricted to employment. They also exert considerable pressure on the city's infrastructure, urban services, housing sector, transport system and the planning of the city. In Bombay, with its 9 million inhabitants, over half are thought to live in squatter settlements or in shanties. The present level of house building in Bombay cannot hope to cope with this problem. It has been estimated that, by the year 2000, 75 per cent of the projected city population of 16 million will live in huts or on the city's pavements. There are acute water shortages and the city faces seemingly insuperable problems in managing refuse disposal. Such conditions have not surprisingly resulted in a high incidence of infectious disease and ill health. The conditions in Bombay are repeated in many other Third World cities struggling to cope with the influx of rural migrants (plate 5.1). In some cases frustration with poor living conditions in the city and the struggle to make a living in the city has led to riots breaking out and to serious disruption of the urban system. For example, in 1984

Plate 5.1   Tegucigalpa, Honduras: migrant housing
Photo: C Hughes

there were food riots in the cities of Tunisia and Morocco and similar riots have occurred in Egypt. Equally, in Iran, frustration at inequalities in urban standards of living has resulted in violence and protest.

International labour migration

Rural–urban migration is only one type of migration flow which occurs in the whole complex of population circulation. Although there is no scope within this chapter to examine in detail the wide variety of types of population mobility or how these change over time (these form the major theme in Hugo's book in this series), it is appropriate to mention one other type of migration flow which has had a very significant impact on the economic development of both regions of origin and destination in Third World countries. During the 1970s and 1980s international migration has become increasingly significant, both in terms of the numbers of people involved and in terms of the variety of origins and destinations. Furthermore it provides for many a new way of making a living in a world where employment opportunities are under increasing pressure. Flows of migrants across international boundaries occur in many parts of the world, often from Third World countries to developed countries. Such flows occur, for example, from several African countries to South Africa, from Mexico to the United States, from North Africa to Europe and from the Caribbean to Britain.

Undoubtedly over the last decade one of the most significant areas of the world importing labour has been the oil-rich countries of the Middle East. The discovery and exploitation of oil in the Gulf economies led in the postwar period to a rapid accumulation of wealth in oil-producing states such as Saudi Arabia, Kuwait and the United Arab Emirates. This was most marked after the oil-price rises of 1973. The accruing wealth led to vast immigration to the oil states, not only of workers seeking employment in the oil industry itself, but of people being drawn into the rapidly expanding labour market of the area. Oil revenues were spent on a wide range of development projects which could not be achieved using indigenous labour. By the early 1980s, 5 million Arab and Asian migrant workers had been drawn to the oil economies in search of a new form of livelihood. This new form of mobility reflects the greater distances over which population distribution has been occurring in the twentieth century as a result of technological advances permitting mass transportation and international travel. Of more importance it illustrates the changing scale of social and economic organization from tribal to national to international labour systems. In a matter of a few decades, nomadic tribal systems in countries like Oman and Saudi Arabia have found their base eroded by the emergence of new social and economic systems. The wage economies resulting from the international oil industry, and its associated urban infrastructure, have resulted in long-distance labour migration and have changed the basis on which these societies are organized.

The demographic impact of international labour flows is very different from that of internal migration, as family regroupment rarely occurs and migrants are forced to return to their home countries at the end of the contract. Similarly, the impact on the economic development of the areas of origin poses its own specific development problems, since migrants not only send home foreign earnings but also transmit some of the technological and cultural characteristics of the foreign societies in which they have worked.

Key ideas

1 Rural–urban migration has stimulated very rapid urban growth in the Third World.
2 The migration process has development implications both for regions of out-migration and for those receiving migrants.
3 In-migration to Third World cities has added to these cities' housing problems and has increased the levels of urban unemployment and under-employment.
4 International labour migration has become of increasing importance in the Third World and has contributed to the emergence of new social and economic systems.

6
Development and
population planning

Earlier chapters have attempted to illustrate the mutual interrelationship of demographic structures, population distribution and economic and social development. The relationships have been shown to be subtle and complex and to vary geographically with rates of population growth and the level of social and economic development within a country or region. All governments design policies, adopt development programmes and pass legislation which directly or indirectly influence population growth or the distribution of a country's population. The *Declaration of Mexico* stressed the view that development policies must reflect the inextricable links between population, resources, environment and development. It also proposed that demographic variables could play an active rather than a passive role in influencing the nature and course of development and that those governments which continued to consider population growth a hindrance to economic development should reconsider their approach to population planning. A distinction is made in this chapter between those policies which indirectly and those which directly affect population composition, growth and distribution.

Social, economic and regional development programmes

Most government policies are not aimed at directly modifying population characteristics, but have an incidental, albeit sometimes very important, influence on population. Clearly a wide variety of social, economic and regional policies exist which are designed to achieve a diverse range of goals but which have a very distinct impact on a country's population distribution

Table 6.1   Population impacts of development programmes

	Population impact		
Development programme	*Employment characteristics*	*Fertility, mortality*	*Migration*
Social programmes			
Education	×	×	–
Status of women	×	×	–
Health	×	×	–
Economic programmes			
Industrialization	×	+	×
Employment policies	×	+	×
Irrigation programmes	+	–	+
Land settlement	+	+	×
Regional Development			
Growth poles	+	–	×
New towns	+	–	×
Improvement of urban housing and infrastructure	–	+	×

Possible impact:
× strong impact
+ some impact
– nature of impact uncertain

and its social characteristics. Table 6.1 summarizes a variety of development programmes which a government might undertake in an attempt to promote specific development objectives in terms of improving the quality of life of all or part of its population. Some of the possible population variables affected by such programmes are indicated, although no evaluation of the 'positive' or 'negative' characteristics of the impacts are made. The types of programmes have been grouped into three broad categories – social, economic and regional programmes. The social programmes involving, for example, compulsory education to a given age, are likely to have some impact on the age of entry to the workforce and also on fertility behaviour, health awareness and thus infant mortality. The impact on migration is less obvious, with contradictory results being reported from different developing countries as to whether educational provision increases or reduces migration.

Economic programmes, such as those encouraging foreign industrial investment and land settlement, are likely to have an impact on population distribution but will have a less clear-cut effect on fertility and mortality trends. Under the regional development section, not surprisingly, the main impacts would be on migration and population distribution, although these might have side effects on other population variables. Specific development projects may therefore have very important implications for population structures or distributions. These changes in turn may create new social and economic environments conducive to changes in mortality or fertility conditions.

Population policies

Population policies are defined here as those policies which seek in a deliberate way to change the size, growth, composition or distribution of a country's population. Frequently conflicts arise between different strands of national legislation, with some policies designed to protect disadvantaged or impoverished groups such as large families, whilst other policies may aim to limit family size. In addition, population policies may have a variety of aims and areas of influence. These may be grouped into three categories: (1) changes in population size, (2) changes in population distribution and (3) changes in population composition.

While considerable differences exist between countries in policies on fertility and migration, most governments are firmly committed to reducing death rates as much as possible. Inversely, policies on fertility vary considerably because of cultural and ideological differences between countries about the right of the state to try to manipulate birth rates, and because of differences in the perceived desirability of increasing or reducing national population size. Direct methods of birth control such as contraception and sterilization were the focus of most of the early family planning programmes. Population planners have become increasingly concerned about ways of motivating people to use these methods and it is certainly the case that marketing has become as important as medicine in promoting birth control. Indirect methods of achieving fertility decline have become increasingly popular. They comprise improving the status of women, raising the legal age of marriage, limitations on family allowances as family size increases, tax-law incentives and increased educational opportunities.

The control of natural population growth has been regarded as the most important form of population planning. However, the development problems resulting from population redistribution, particularly in the capital cities of the Third World, have also led some governments to attempt to initiate policies designed to change population distribution. In a few cases rural migrants have been transported back to the areas from which they came, or strict controls on population movements, by means of registration schemes, have been introduced. Most policies on population distribution, however, operate by indirect rather than direct measures. They are intended to encourage migrants either to stay in their home areas or else to

一对夫妇只生一个孩子

Plates 6.1, 6.2 and 6.3   Chinese family planning. The caption above reads 'One Family, One Child'.   *Photos*: A J Jowett

encourage them to move to new towns, new industrial zones or to work on specific agricultural schemes.

Policies to alter the composition of the population are particularly hard to implement and, because they are often ethically unacceptable, they are seldom publicized. Migration may change the age, sex or race composition of the population of a given area. At the scale of international migration this is particularly true, with many countries operating strict controls on immigration and family regroupment. Having indicated that a wide range of population policies have development implications, the remainder of this chapter focuses on the policy tools and effectiveness of just one form of intervention: family planning.

An assessment of family planning policies

Birth control and family planning programmes have been operated by many Third World countries in the post-war period, but the nature of these programmes has changed through time. The first impetus for such policies came in Asia where high birth rates gave cause for alarm in an already densely populated zone. As early as 1952 the Indian government adopted a policy of family planning; today almost every Asian country has a family planning programme of incentives and disincentives to influence individual choices on the number and spacing of children in each family. Incentive schemes vary. In Bangladesh female participants in the programme receive a new sari. By contrast, in China, those willing to sign the one-child pledge receive cash payments. Sri Lanka operates a scheme whereby those with only two children receive pension benefits. In one Indian state, Madhya Pradesh, families receive rewards for spacing their children at four-year intervals. In addition to rewards for participating in birth control schemes, social rewards are also offered to those who limit their family size. Families may receive free or preferential medical treatment, as in Korea. They may also be eligible for preferential assistance with housing. In Singapore children from two-child families, where one parent has been sterilized, have priority in schooling. Disincentives also exist in many countries, with extra tax being paid by families of a certain size or with fourth and subsequent children not being eligible for assistance with schooling.

The schemes outlined above are all based on individual decision-making and choice. Some countries such as India, Indonesia, Bangladesh and Thailand have also experimented with community-based schemes. In these schemes the idea is that a sense of social responsibility will be generated and that villages will receive rewards for their family planning record. These rewards usually take the form of investment in development projects to create further employment.

So far the examples quoted have been from Asia where family planning has been implemented on the largest scale. In Latin America and the

Case study E

An evaluation of China's population policies, 1949–85
by A. J. Jowett

During thirty-five years of Communist rule the population of China has increased by about 500 million, from 541 million in 1949 to 1036 million at the end of 1984. This moderately rapid growth of population, which has averaged 2 per cent per year, gave China an annual increase of more than 23 million in 1970, and at that time a population of Scotland's dimension was being generated every three months of the year. A plot of annual birth rates, death rates and population totals indicates the fluctuating fortunes of China's population since 1949 (figure E.1), and reveals that the two most outstanding features are the major disasters centred on the famines of 1960/61 and the spectacular decline of fertility in the 1970s.

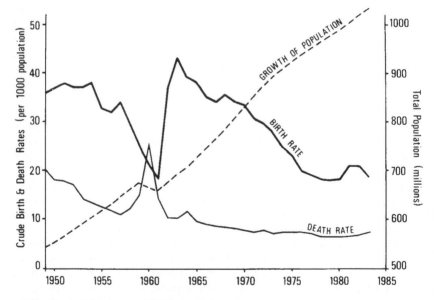

Figure E.1   China: demographic trends
Source: A J Jowett, 1986, China: population change and population control, *GeoJournal*, 12, 351 (reproduced by permission)

Case study E (*continued*)

Population policies since 1949

During the 1950s and 1960s China's leaders oscillated between pro- and anti-natalist policies before finally opting for an intensive programme of birth control in the 1970s in the aftermath of the Cultural Revolution (1966/ 69). So successful has this been that, despite a population increase of about 150 million in the 1970s, the Chinese managed 10 million fewer births in 1979 than had been the case in 1971, the birth rate was halved and the natural growth rate of population declined from 25.83 per 1000 to 11.61 (figure E.1). Such a rapid and radical transformation of attitudes with regard to fertility in so large, so poor and so rural a country as China is quite remarkable.

China's spectacular achievement in reducing fertility, though dependent on a wide range of variables, appears to owe much to the government-sponsored family planning programme. In the 1970s the reduction of fertility became a national priority pursued with consistency and vigour through a wide range of policies. The government spelt out its three reproductive norms in the slogan *wan-xi-shao* (later, longer, fewer), meaning late marriage, a long interval between births and few children. In the early 1970s 'few' meant two children in urban areas and three in the countryside, but in 1977 the target was lowered to two throughout China and the expectation was that a four-year interval would elapse between the two births. In pursuit of these objectives an all-embracing administrative structure was established for the delivery of family planning information, targets and supplies. The effectiveness of the Chinese birth planning programme stems primarily from its unique organization which can effectively transmit and enforce the policies of central government at local level.

In contrast to most other countries China has put its major family planning emphasis on the use of intra-uterine devices (IUDs), so much so that some 65–70 per cent of all the world's IUD users are to be found in China. Overall some 70 per cent of married couples of child-bearing age are currently practising contraception. Thus the family planning campaign of the 1970s raised the contraceptive use in China to the levels presently experienced in the developed world. Even so, induced abortion is also extensively used to prevent unwanted births.

Despite all the success achieved in the 1970s in the field of fertility control, the government decided that an even more stringent population

Case study E (*continued*)

control policy was required. Thus in 1979 the leadership put forward a policy of 'one couple, one child' and at that stage China entered a new phase of its birth planning campaign. In an attempt to establish the one-child family as the norm in China, substantial financial incentives and disincentives were introduced into the family planning programme. The basic offer to parents in 1979 was a 10 per cent salary bonus for limiting families to one child and a 10 per cent salary reduction for those who produced more than two. These financial economic sanctions were to operate for the first 14–15 years of the child's life. The most recent phase of the population control programme involves a much more stringent implementation of the one-child norm. Penalties previously levied on those who gave birth to a third child are now to apply to those who give birth to a second child. Such an ambitious programme, it is hoped, will restrict total population to within the targeted figure of 1200 million in the year 2000.

The promotion of the one child policy represents China's most definitive shift from a pro-natalist to an anti-natalist line, from an optimistic view of people as producers (hands to work) to a more pessimistic view of people as consumers (mouths to feed). The problems China faces in feeding, clothing, housing, educating and employing the population are increasingly being blamed, rightly or wrongly, on the detrimental impact of a large and rapidly expanding population.

Important though the government-sponsored family planning programme was in reducing fertility in China, it needs to be stressed that it was but one element in a much broader population control programme. Indeed part of the reason for the success of the family planning programme was that by the 1970s Chinese society had changed significantly. Improvements in education and health, income and the changing role of women in society combined to make a very positive contribution to the goal of reducing fertility.

Education for women, for example, is one of the major factors which generates a reduction in fertility. China conforms to this pattern for in 1981, while illiterate mothers averaged 4.7 births, those with a senior secondary education averaged 2.4 and mothers with a tertiary education were producing less than two children. The age at which women marry also has an important influence on the rate of population growth. Early marriage is associated with high fertility and late marriage with low fertility. Later marriage certainly played an important role in lowering the birth rate in China. The average age of females at first marriage rose from 18.5 years in

Case study E (*continued*)

1949 to about 23 years in 1979, with the increase being particularly rapid in the late 1970s.

Despite the disaster of 1960, the pace of demographic modernization in China has been most impressive. As seen in figure E.2, China's level of fertility and life expectancy is totally out of character with her low level of

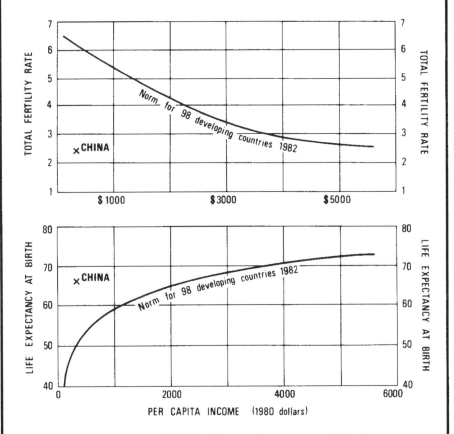

Figure E.2 China and the Third World: per capita income total fertility rates and life expectancy at birth
Source: A J Jowett, 1986, China: population change and population control, *GeoJournal*, 12,361 (reproduced by permission)

Case study E (*continued*)

economic development, as measured by the traditional criteria of per capita income. Life expectancy is some fifteen to twenty years greater than would be expected for a country at China's level of per capita income. Indeed, to achieve China's level of life expectancy, one would expect a per capita income almost ten times greater than currently prevails in China. The situation is even more impressive in the case of fertility levels within China (figure E.2).

This case study appears to support the belief that what appears to control the pace of demographic modernization is not so much the overall level of economic development attained by a country but the widespread participation of the population in the developmental process. China has, within a Third World context, an impressive record in ensuring a widespread and equitable distribution of food supplies, health care and education; in improving the status of women; in providing career opportunities independent of child-bearing and in ensuring the access of all, especially the poor, to family planning services. These achievements, allied to an intense and, since 1970, a sustained commitment on the part of government to a reduction in the rate of population growth, have given China advanced levels of demographic modernization at an early stage of economic development.

Caribbean population planning has been regarded as much more controversial and most countries have not introduced policies to reduce fertility. Although these countries do not suffer from the same problems of 'over population' in rural areas, they do experience considerable pressure on resources in the urban agglomerations. Where family planning programmes do exist, they tend to be part of the public health programmes but are generally not high profile and do not comprise the same range of incentive schemes. Columbia has one of the most advanced programmes in Latin America. Other countries concerned to reduce population growth are the central American countries such as Costa Rica and also Mexico.

The African countries were slow to become actively involved in population planning, although they have not experienced the same resistance as those of Latin America. Ghana launched a family planning programme in 1970 when it aimed to reach 10 per cent of women in the fertile age group within the first five years of the programme. Tunisia has been one of the countries most favourably disposed towards population control in the Arab world. In 1973 President Bourguiba publicly stated that

Table 6.2   Population planning policy and fertility decline

	Government policy		
Relative level of socio-economic development	*Policy with specific objective of reducing fertility*	*Policy with no specific objective of reducing fertility*	*No policy or negative policy*
High	COLUMBIA DOMINICAN REPUBLIC HONG KONG MALAYSIA MEXICO PHILIPPINES KOREA SRI LANKA TURKEY	BRAZIL CHILE CUBA ECUADOR Iraq PERU VENEZUELA	ARGENTINA N. KOREA Syria
Medium	CHINA EGYPT Ghana GUATEMALA INDIA INDONESIA IRAN Kenya MOROCCO Pakistan THAILAND TUNISIA VIETNAM	Algeria S. Africa Zaire Zambia Zimbabwe	Bolivia Burma Cameroon Madagascar Saudi Arabia
Low	Bangladesh Nepal Senegal	Afghanistan Haiti Mali Nigeria Sudan Tanzania Uganda	Angola Ethiopia Guinea Ivory Coast Kampuchea Malawi Mozambique Niger Upper Volta Yemen AR

Note: Developing countries with a reduction (1960–77) in the crude birth rate of less than 10 per cent are in lower case, and those with a reduction of more than 10 per cent are in upper case.

sterilization was not in conflict with Islamic law. Tunisia has adopted a family planning policy as well as attempting to improve the status of women. Child allowances are also limited to the first four children and the legal age of marriage has been raised. Other Arab countries such as Morocco and Egypt also have family planning programmes.

The relationship between family planning programmes and declining crude birth rates may be seen in table 6.2. In the table countries are grouped by level of socio-economic development and cross-classified by whether they have specific policies to reduce fertility, policies with no specific objective to reduce fertility or no policies at all. It may be observed that most of the Asian countries do have a policy with the specific objective of reducing fertility, whilst most of the Latin American and African countries are grouped in the second and third columns.

In the table countries which actually achieved a reduction in the crude birth rate of 10 per cent between 1960 and 1977 are printed in capitals. Two trends are clear. Firstly, a high proportion of the countries with specific policies to reduce fertility have also experienced a lowering of the crude birth rate, whilst this has not occurred in the other two columns. Secondly, most countries with only low levels of socio-economic development have no fertility policies or have policies which do not seek to reduce fertility. Of particular interest is the fact that most of the countries in the medium level of socio-economic development with a specific policy to reduce fertility actually experienced a marked fertility reduction. Amongst the few countries in the low socio-economic development group which implemented policies to reduce the birth rate none experienced a substantial fertility reduction. This table underlines the intimate two-way relationship between development levels and fertility and also shows how population planning can be viewed as only one component in the population–development relationship. Undoubtedly the countries which have adopted the most comprehensive approaches to family planning and have operated them within an appropriate social and economic environment have also been the most successful. For example, in South Korea, when the family planning programme was reformulated in 1981 with extra incentives being offered to acceptors of sterilization, it was found that a dramatic rise in the number of acceptors resulted. In Sri Lanka policies were similarly adapted to attract self-employed agriculturalists and an increase in sterilization acceptors also took place.

Family planning policies, while they may appear to be modestly effective in reducing rates of population growth, do not remove the problems of under-development. It should be noted that the promotion of family planning policies in Third World countries by international agencies financed by western governments has been treated with increasing scepticism. The cry of the more developed countries to the Third World in the 1960s to conserve resources and to control population growth had a touch of the Mad Hatter's tea party ' "No room! No room!" they cried out when they saw Alice coming. "There's plenty of room!" said Alice indignantly.'

Representatives of the Third World have quite rightly responded to western suggestions that family planning is the panacea to their problems by arguing that inequality in the international trade of raw materials, manufactured goods, finance and technology is the root cause of their difficulties. As Ehrlich (see page 1) has pointed out, each American child absorbs fifty times as many resources as an Indian child. Solutions to the problems of population and development should be concerned as much with resource management issues in the more developed countries as with family planning programmes in the Third World.

Fortunately criticism of the motivations underpinning the early family planning programmes of the 1950s and 1960s has not led to an abandonment of all population policies but rather to a restructuring and reorientation of family planning programmes and to their integration into development strategies. In place of policies imposed by international agencies, with their moderately unsuccessful record, there has been an increased emphasis on indigenous programmes adapted to the social and cultural attitudes of the populations involved. Programmes of this sort have much greater potential for increasing the quality of life of the populations in which they are applied, because they arise out of a public consensus that the demographic behaviour of the individual can have a very real impact on the welfare of the community at large.

Key ideas

1 Most government policies, whether they are intended to or not, affect population.
2 The focus of population planning in the Third World has shifted from policies which directly encourage the adoption of birth-control measures to policies which are designed to favour fertility decline indirectly.
3 Countries with the lowest levels of socio-economic development have not only been less likely to adopt family planning policies but also less successful in applying them.

References, further reading and review questions

Items for further reading are indicated with an asterisk.

Preface and chapter 1

*Brandt, W. *et. al.* (1980) *North–South: A Programme for Survival: Report of the Independent Commission on International Development Issues*, London, Pan.
Declaration of Mexico (1984) 'People', 11, 10.
*Ehrlich, P. (1971) *The Population Bomb*, London, Pan.
*Ehrlich, P. and Ehrlich, A. (1972) *Population, Resources, Environment*, San Francisco, Freeman.
Meadows, D. *et. al.* (1972) *The Limits to Growth*, New York, Universe Books.
Population Reports (1983) *Migration, Population Growth and Development* (Series M7, Population Information Programme), Baltimore, Johns Hopkins University.
*World Bank (1985) *The World Bank Atlas 1985*, Washington, World Bank.

1 Compare and contrast the attitudes to population growth taken at the Bucharest and Mexico World Population Conferences (see preface).
2 Describe the patterns evident in figure 1.2 and explore the advantages and disadvantages of using gross national product per capita as a measure of economic development.

3 Attempt an explanation of the differences between the British, Brazilian and Kenyan population pyramids.

Chapter 2

Bongaarts, J. (1978) 'A framework for analysing the proximate determinants of fertility', *Population and Development Review*, 4, 105–32.
Bongaarts, J. (1985) 'The fertility inhibiting effects of the intermediate fertility variables', in F. Shorter and H. Zurayk (eds) *Population Factors in Development Planning in the Middle East*, Cairo, Population Council, 152–69.
*Caldwell, J. (1982) *Theory of Fertility Decline*, London, Academic Press.
*Cleland, J. and Hobcraft, J. (eds) 1985 *Reproductive Change in Developing Countries*, Oxford, OUP.
*Jones, H. (1981) *A Population Geography*, London, Harper & Row.
Kchir, S. (1979) *La Mortalité Infantile et Juvenile dans le Grand Tunis*, Tunis, Cert. Aptitude de Recherche, Université de Tunis.
Lardinois, R. (1982) 'Une conjoncture de crise démographique en Inde du sud au 19e siècle', *Population*, 37, 371–405.
Population Reports (1985) *Fertility and Family Planning Surveys: an Update* (Series M8, Population Information Programme), Baltimore, Johns Hopkins University.
*Rodgers, G. (1984) *Population and Poverty*, Geneva, ILO.

1 How are crude birth and death rates measured? Why is it difficult to compare population trends in different countries using these measures?
2 Why have death rates in developing countries not fallen to the same levels as those in the more developed economies?
3 What light has the World Fertility Survey thrown on the determinants of fertility decline in the Third World?

Chapter 3

Caldwell, J. (1982) *Theory of Fertility Decline*, London, Academic Press.
Chung, R. (1970) 'Space-time diffusion of the transition model', in G. Demko *et al.* (eds), *Population Geography: A Reader*, New York, McGraw Hill, 220–39.
*Clarke, J. (1985) 'Islamic populations: limited demographic transition', *Geography*, 70, 118–28.
Davis, K. (1963) 'The theory of change and response in demographic history', *Population Index*, 29, 345–66.
*Easterlin, R. (1980) *Population and Economic Change in Developing Countries*, Chicago, University of Chicago Press.

Notestein, F. (1945) 'Population: the long view', in T. Schultz (ed.) *Food for the World*, Chicago, University of Chicago Press, 36–57.

*Woods, R. (1982) *Theoretical Population Geography*, London, Longman.

*World Bank (1985) *Population Change and Economic Development*, Oxford, OUP.

1 Explain what you understand by the term 'demographic transition theory'.
2 How does Caldwell's work help in understanding the mechanisms of Third World fertility decline?
3 If you were an economic planner in an African or South Asian country, what economic and social policies would you introduce if you wished indirectly to reduce fertility levels?

Chapter 4

Boserup, E. (1965) *The Conditions of Agricultural Growth*, London, Allen & Unwin.

Boserup, E. (1980) *Population and Technology*, Oxford, Blackwell.

*Harrison, P. (1980) *The Third World Tomorrow*, Harmondsworth, Penguin.

Mooley, D. (1981) 'Large scale droughts over India, 1871–1978', *Annals of the National Association of Geographers of India*, 1, 17–26.

*Simon, J. (1981) *The Ultimate Resource*, Oxford, Martin Robertson.

*Visaria, P. and Visaria, L. (1981) 'India's population: second and growing', *Population Bulletin*, 36, 4.

*World Bank (1985) *Population and Economic Development*, Oxford, OUP.

1 Review the circumstances which surrounded the Ethiopian famine of 1984/85 and evaluate the extent to which the situation reflects a Malthusian crisis.
2 From an examination of figure C.1 (page 48) describe the steps which India has taken to feed its rapidly growing population.

Chapter 5

Findlay, A. (1982) 'Migration planning', *Applied Geography*, 2, 221–30.

Gilbert, A. and Gugler, J. (1982) *Cities, Poverty and Development*, Oxford, OUP.

Hugo, G. (1979) 'The impact of migration on villages in Java', in R. Pryor (ed.) *Migration and Development in S.E. Asia*, Oxford, OUP, 204–11.

Mabogunje, A. (1970) 'Systems approach to a theory of rural–urban migration', *Geographical Analysis*, 2, 1–18.
*Ogden, P. (1984) *Migration and Geographical Change*, Cambridge, CUP, 63–71.
*Prothero, R. M. and Chapman, M. (eds) (1985) *Circulation in Third World Countries*, London, Routledge & Kegan Paul.
*Pryor, R. (ed.) (1979) *Migration and Development in S.E. Asia*, Oxford, OUP.
Wilheim, J. (1984) 'Sao Paulo', *Cities*, 1, 538–42.

1 What have been the factors favouring population migration to Third World cities?
2 In what sense is migration a selective process, and what are the development consequences of this selectivity?
3 Why does migration present many problems for the town planner of a Third World city?

Chapter 6

*Gould, W. and Lawton, R. (eds) (1986) *Planning for Population Change*, London, Croom Helm.
Jowett, A. J. (1986) 'China: population change and population control', *GeoJournal*, 12, 349–63.
*Nortman, D. and Hofstatter, E. (1976) *Population and Family Planning Programs*, New York, Population Council.
*Population Reports (1985) *Fertility and Family Planning Surveys: an Update* (Series M8, Population Information Programme), Baltimore, Johns Hopkins University.

1 What factors would you take into account in formulating a family planning programme?
2 How effective have family planning programmes been in the Third World?
3 Evaluate China's population policies since 1949.

Index